Breal [illegible] Mold

Tanks in the Cities

Kendall D. Gott

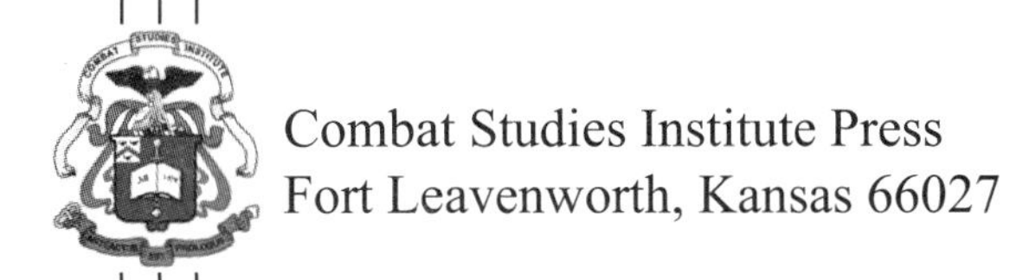

Combat Studies Institute Press
Fort Leavenworth, Kansas 66027

Library of Congress Cataloging-in-Publication Data

Gott, Kendall D.
Breaking the mold : tanks in the cities / Kendall D. Gott.
p. cm.
Includes bibliographical references and index.
1. Street fighting (Military science) 2. Urban warfare. 3. Tanks (Military science). 4. Armored vehicles, Military. I. Title.
U167.5.S7G59 2006
355.4'26--dc22
2006016027

For sale by the Superintendent of Documents, US Government Printing Office
Internet: bookstore.gpo.gov Phone: toll free (866) 512-1800; DC area (202) 512-1800
Fax: (202) 512-2250 Mail: Stop SSOP, Washington, DC 20402-0001

ISBN-10 0-16-076223-5
ISBN-13 978-0-16-076223-9

Foreword

Few lessons are as prevalent in military history as is the adage that tanks don't perform well in cities. The notion of deliberately committing tanks to urban combat is anathema to most. In *Breaking the Mold: Tanks in the Cities*, Mr. Ken Gott disproves that notion with a timely series of five case studies from World War II to the present war in Iraq.

This is not a parochial or triumphant study. These cases demonstrate that tanks must do more than merely "arrive" on the battlefield to be successful in urban combat. From Aachen in 1944 to Fallujah in 2004, the absolute need for specialized training and the use of combined arms at the lowest tactical levels are two of the most salient lessons that emerge from this study. When properly employed, well-trained and well-supported units led by tanks are decisive in urban combat. The reverse is also true. Chechen rebels taught the Russian army and the world a brutal lesson in Grozny about what happens when armored units are poorly led, poorly trained, and cavalierly employed in a city.

The case studies in this monograph are high-intensity battles in conflicts ranging from limited interventions to major combat operations. It would be wrong to use them to argue for the use of tanks in every urban situation. As the intensity of the operation decreases, the second and third order effects of using tanks in cities can begin to outweigh their utility. The damage to infrastructure caused by their sheer weight and size is just one example of what can make tanks unsuitable for every mission. Even during peace operations, however, the ability to employ tanks and other heavy armored vehicles quickly can be crucial. A study on the utility of tanks in peace operations is warranted, and planned.

Breaking the Mold provides an up-to-date analysis of the utility of tanks and heavy armored forces in urban combat. If the recent past is a guide, the US Army will increasingly conduct combat operations in urban terrain, and it will therefore be necessary to understand what it takes to employ tanks to achieve success in that battlefield environment. *CSI—The Past is Prologue!*

Timothy R. Reese
Colonel, Armor
Director, Combat Studies Institute

Preface

This work examines the use of tanks in urban warfare. It seeks to provide insight and a historical precedence on the wisdom of employing tanks in an inherently dangerous dimension of the modern battlefield, intensifying the shortcomings in technological design and the lack of crew training for city fighting. Instead of being a legacy system ready for the scrap heap, tanks are still a vital component of the US Army, even in the streets.

During most of my Army career from 1978 to 2000, I either served in or supported armor or mechanized units. This may or may not make me a subject matter expert, but the topic is very familiar to me. As an M60A3 tank platoon leader, I witnessed firsthand the US Army's doctrine and attitude for using armor in the city—it just wasn't to be done. During my three tours in Germany, armor units spent a great deal of time in the forested hills overlooking picturesque valleys, but never deployed to towns and villages. Naturally, the specter of maneuver damage by heavy vehicles had something to do with this, but even the general defense plans had no serious discussion on armor in urban fights, presumably leaving such operations to the infantry or the Germans. Faced with the narrow maze of streets in most German towns, even the tankers thought their place was in the countryside. There were enough experiences with gun tubes rammed through the sides of buildings, crushed civilian vehicles, and crawling convoys in peacetime to give any tanker pause before choosing a city as a battlefield. Tanks were made to go fast and shoot far; we could not do that in downtown Fulda or Frankfurt.

The Israeli experiences in the 1973 War reinforced these attitudes. After the Israelis' stunning victory in the desert, many of us tried to emulate their procedures and tactics. Like the Israelis, tank commanders were encouraged to fight with their hatches open for rapid target acquisition. Most of us took stock that the Israelis had used American tanks against the Soviet-made counterparts. Again, most of us in the armor corps were thinking in conventional ways and did not apply the Israeli experience to the city. In fact, the Israeli debacle in Suez City only reinforced our worse fears.

Since my tank and scout platoon days, a number of world events have illustrated the dynamics of successful employment of armor in urban terrain. If I doubted the wisdom of using tanks in the city in my youth, I am a firm believer now. The "Combat Arm of Decision" is just as relevant, if not more so, than ever before. Obviously, those in various levels of leadership around the world must think so too, as they have broken the mold and sent tanks into the urban fray.

In this work, I have endeavored to use a narrow focus in the roles and functions of the supporting arms. This is not meant to be a "how to fight" manual, as the US Army and the other branches of service have published doctrine on the subject. I leave it to the readers to seek out those documents. Research included no classified material. Defending units are in italics to avoid confusion. Because accurate battle losses were often difficult to obtain, the best approximations are used. The footnotes highlight particularly noteworthy sources should the reader wish to seek further information. Participants and historians are still analyzing the battle for Fallujah; therefore, source material is lean. No doubt, more information will become known in the coming years. It is not too early though to glean useful insights to the use of armor from that instance for consideration.

My thanks go to Colonel Timothy Reese, Director, Combat Studies Institute and to the topic and editorial board for their support and input. In addition, thanks go to Mr. John McGrath and Mr. Matt Matthews for research materials and their expertise on the subject. No author is complete without an editor, and I thank Betty Weigand for her efforts in making this work a reality.

The views expressed in this publication are mine and do not necessarily represent the official policy or position of the Department of the Army or the Department of Defense.

Ken Gott
Combat Studies Institute

Contents

Page

Page

Maps

Introduction

During World War I, the tank was developed as an infantry support weapon to exploit breaches made in enemy lines. Technological limitations in speed, range, and mechanical reliability kept tank doctrine at the tactical level until the German offensives in 1939–40 showed that modern armored forces were a key element to the operational level of warfare. Yet, there was virtually no discussion of employing armor in the cities. Even famed military historian and early theorist of modern armored warfare John Frederick Charles Fuller seldom mentioned using tanks in urban terrain, and then only to dissuade their use. Avoiding the employment of armor in cities is a long-held trend that holds sway in most modern armies. Historically, battles for large cities are full of examples of high casualties and massive collateral damage, and the specter of a tank's easy destruction in the close confines of urban terrain weighs heavily on commanders and military planners. However, in a historical context, the vulnerability of armor in cities is proven to be overestimated and outweighed by the ability of the tank to bring its heavy firepower to the urban fight.[1]

Military operations on urbanized terrain (MOUT) are not new to the US Army. World War II has numerous examples of US military personnel fighting in cities. What is new is the increasing use of tanks and other armored combat vehicles in cities. What was once considered taboo is now becoming commonplace because of the worldwide demographic shift of rural populations to cities. Some analysts estimate that by 2010 over 75 percent of the world's population will live in urban areas, thus shifting the future battlefields to within their limits. Additionally, the requirement to conduct stability and support operations will require the occupation of cities, whether large or small. Future military leaders will not have the luxury of avoiding Sun Tzu's axiom, "The worse policy is to attack cities. . . . Attack cities only when there is no alternative."[2]

Urban operations will become a necessity in the future because of these trends. To defeat an enemy, his major urban centers must be seized as they increasingly represent the power and wealth of a nation. This is because cities not only seat the ruling government, but also hold the industrial base, transportation network, and the heart of the country's economic and cultural centers.

Future battles for cities will be fraught with the same perils that made armies of the past avoid them. Narrow streets are ideal ambush sites, and the risk of high casualties is great. Rarely is there a swift and sure outcome. In cities, the enemy often chooses to mix with the civilian populace. Heavy firepower is often counterproductive as the resulting rubble makes fighting positions even more formidable. Collateral damage will kill or wound civilians, while various media beam the pictures and tales of their suffering across the globe. Logistics and medical evacuation are difficult at best. Thus, it is not surprising military leaders prefer to give cities a wide berth.

During World War I, England and France developed tanks concurrently for a single specific purpose. Tanks were to beat a path for infantry in frontal attacks against an entrenched enemy with rifles and machine guns. By design, tanks were an infantry support weapon armed with machine guns and light cannons and equipped with enough armor protection to close with the enemy unscathed. Technological limitations and mechanical reliability severely restricted the use and effectiveness of tanks. In World War II, the tank was still an infantry support weapon, and the early nature of the conflict was traditional in that major urban centers were generally bypassed. This changed in 1941 as the Soviets adopted the tactic of holding on to their large cities and forcing the Germans to attack into them. The epic battle of Stalingrad is one example that cost the Germans dearly in manpower and armored vehicles. In 1943–44, it was America's turn to learn firsthand the horrors of urban combat on a large scale as US soldiers advanced through Italy and France and into Germany. In these offensives, the Germans chose to fight from the cities, forcing the Americans to attack into them. The lessons were grim and reinforced the axiom that it was far better to avoid city fighting if possible. This is reflected by tank design toward the end of the war as it departed from the infantry support role. By then all sides of the conflict developed and fielded tanks suited for tank-on-tank fights. This emphasis continued unabated for 60 years, although the role and effect of armor on the battlefield has come under scrutiny from time to time.[3]

Doctrinally, the US Army has tried to come to grips with urban warfare and the employment of armored forces in that spectrum of battle. Field Manual 90-10, *Operations in Urbanized Terrain,* for decades served as a blueprint for how to fight in such terrain. In 2003, Field Manual 3-06, *Urban Operations,* replaced Field Manual 90-10, incorporating the experience gained during the Iraq War. Various supplemental documents and a myriad of professional periodicals address this subject as well. The vast majority of documents on the subject of urban warfare is quite candid about the trials faced in this type of fighting and advocate avoiding the battle if possible. Generally, armored forces are relegated to the supporting role while the infantry remains dominant. This doctrine is time-tested with battlefield experience and through exercises at the Joint Readiness Training Center at Fort Polk, Louisiana; the Combat Maneuver Training Center at Hohenfels, Germany; and the Mounted Urban Combat Training Site at Fort Knox, Kentucky.

With the rise of insurgent wars and the proliferation of advanced weaponry, many American defense planners virtually wrote off the tank as a legacy system in need of retirement. In the 1990s, the US Army slashed its armor force and focused on developing and fielding vehicles and units that could rapidly deploy but not necessarily fight as effectively. The 2003 Iraq War and the subsequent occupation have shown that rather than being a mobile coffin, heavy armor still provides an overwhelming and lethal capability to any force. Its firepower, mobility, and shock combine to defeat an enemy, even those armed with modern antitank weapons. There are drawbacks though. The tank's heavy weight and large size often restricts its speed and areas of use. Blind spots and frequently restricted turret traverse and elevation hinder the tank's substantial firepower. Historical examples and recent experience show, however, that when employing armor in combination with infantry, supporting artillery, and air power, the tank is a dominant player in urban warfare.[4]

The representative case studies in the next five chapters show the evolution of the use of tanks in urban warfare. In the early case studies, the tanks were employed almost as an afterthought. Later case studies show tanks as an increasingly integral part of the operation plan. In each case, tanks proved their ability and worth, and when properly employed with supporting arms, tanks remain a relevant, if not vital, weapon on future urban battlefields.

Notes

1. John F.C. Fuller, *Armored Warfare* (Westport, CN: Greenwood Press, 1994), 44–48. Fuller saw tanks as an open-country weapon used to achieve operational-level results. Although the use of armor in this role was revolutionary, Fuller was still quite conventional in avoiding battle in large cities.

2. Department of the Army, Field Manual (FM) 3-06, *Urban Operations* (Washington, DC: US Government Printing Office, 2003). Sun Tzu, *Art of War*, trans. Samuel B. Griffith (New York: Oxford University Press, 1963), 78.

3. Trevor N. Dupuy, *The Evolution of Weapons and Warfare* (Fairfax, VA: Hero Books, 1984), 221–222. See also Richard Simpkin, *Tank Warfare: An Analysis of Soviet and NATO Tank Philosophy* (London: Brassey's Publishers, 1979), 164. Simpkin gives scant attention to urban warfare within his 250-page book. He points out that urban combat is inevitable, but does not present an argument for employing armor in cities.

4. Patrick Wright, *Tank: The Progress of a Monstrous War Machine* (New York: Penguin Putnam, Inc., 2002), 429–431. The M1127 Stryker Infantry Vehicle is an example. This light vehicle was capable of rapid deployment, but received expedient armor upgrades to make it battle-worthy against such legacy systems as the RPG-7.

Chapter 1

Sherman Tanks in the Streets: Aachen, 1944

In 1939, the US Army had less than 400 armored vehicles, many of obsolete design, divided between the Infantry and Cavalry branches. At that time, there was not yet a separate branch of Armor. The German victories over Poland and France in 1940, spearheaded by panzer forces, shocked the American military. In response, the US Army commenced a rapid program to manufacture and field tanks. The pace of the war and the need to raise and equip massed forces prevented the rapid deployment of suitable armored vehicles. As a result, American armor did not fight its first major battle until February 1943 in Tunisia. There the inexperienced 1st Armored Division was roughly handled by the Germans but managed to come back fighting. Thereafter, with the exception of the short campaign in Sicily, American armor was not able to demonstrate its real potential for 18 months due to the harsh terrain in Italy and the need to conserve resources for the cross-channel invasion.[1]

By 1944, the armored division became a balanced force with a flexible command system. This consisted of three combat commands (A, B, and Reserve) that were allocated divisional resources according to the tactical situation. These divisional resources typically consisted of three tank battalions, three armored infantry battalions, three self-propelled artillery battalions, a tank destroyer battalion, an armored cavalry squadron, an engineer battalion, and divisional services. There were also some 40 independent tank battalions serving in Europe. Although generally allocated to provide close support for infantry divisions, they occasionally operated as semi-independent armored groups.

Doctrine for employing these armored formations was embryonic and constantly changing. This evolving US doctrine was generally conservative in approach and based on operating in open country. Only a few lines of doctrine were devoted to employing armored forces in urban terrain, and it was strictly to support the infantry effort. In fact, the infantry manuals at the time gave scant attention to urban terrain, mentioning villages and towns but not major cities. Commanders were urged to avoid fighting in built-up areas and encouraged to bypass them whenever possible. Tactics advocated in the doctrinal manuals included a methodical firepower-based approach. Advancing units were committed in a decentralized manner out of necessity and cleared strongpoints by a series of bounds. Frequent halts to reestablish contact with adjacent units were needed to prevent fratricide. Tanks were, by design, infantry support weapons and would accompany the riflemen, lending their heavy weapons against enemy strongpoints.[2]

The greatest weakness of the US armor doctrine was the weapons systems themselves. During World War II, the primary US Army tank was the M4 Sherman. Weighing approximately 35 tons, it mounted a 75mm general-purpose gun firing high explosive armor-piercing and white phosphorus rounds. The tank had a reputation for mechanical reliability, which was its best attribute. The Sherman was designed as an infantry support tank. To deal with enemy armor, the Americans later developed the M10 tank destroyer, which was based on the M4 chassis but mounted a 75mm high-velocity gun. This gun could penetrate most German armor and was adept at dealing with thick walls and fortifications. Its armor was even thinner than the Sherman's armor though, so it could dish out far more than it could take. Both of these armored vehicles were nine feet wide and able to maneuver in most of the narrow streets of Europe.

When American armor was employed, it was evident to the crews that they were heavily outgunned by the German panzers, which also had superior armor protection. Both the M4 Sherman and M10 tank destroyer were extremely vulnerable to German tank fire and to the wide assortment of enemy antitank weapons, including the *panzerfaust*, which was a hand-held single-shot recoilless weapon firing a shaped charge. Although the *panzerfaust* had a very short range of 30 meters, it could devastate American armor. The Germans produced huge numbers of these cheap and effective weapons, and American tank crews learned quickly to avoid both the German panzers and infantrymen armed with the dreaded *panzerfausts*.[3]

The Americans generally overcame the deficiencies of their armor by fielding a large number of tanks and by using massed artillery and aerial firepower. After the breakout from the Normandy hedgerows from August to September, the Allied armies raced across France after the retreating Germans. As the Allies neared the German border, the critical supply situation slowed and occasionally even halted the advance. The logistical situation restricted ammunition expenditure and movement until the end of 1944 with the capture of the port of Antwerp. By the time the Allies approached the border of Germany, their armies were husbanding their logistics and poor weather conditions hampered air operations. An Allied pause gave the Germans the opportunity to recover enough to form a stubborn defense of their homeland. The period of massed Allied armies driving across open countryside was over.[4] (See Map 1.)

The Westwall

The Siegfried Line or *Westwall* as the Germans called it ran from the Dutch frontier to the Swiss border and guarded Germany from an invasion from the west. These defenses consisted of a series of obstacles and

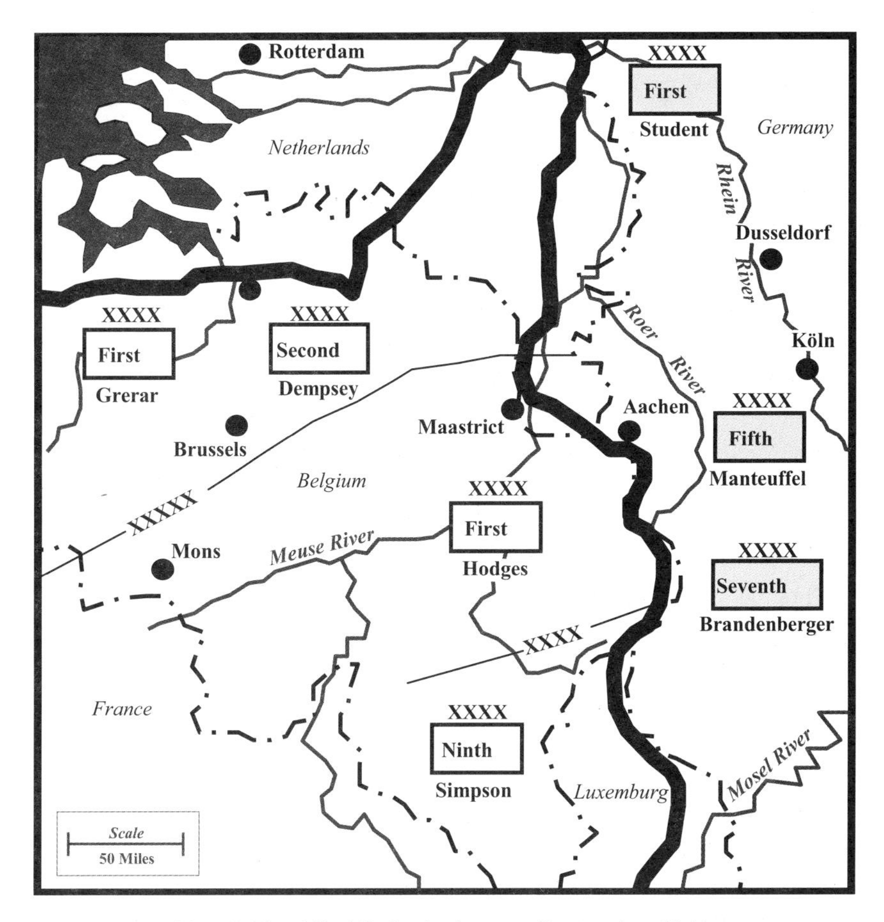

Map 1. The Allied limit of advance, September 1944.

fortifications backed by mobile reserves and artillery to contain and destroy local penetrations. Construction on the line began in 1936 after the Germans reoccupied the Rheinland, but had languished in 1940 after the fall of France. During the previous four years, the defenses were a bit dilapidated but still a significant force multiplier. Unfortunately for the Germans, many of the troops, tanks, and guns needed to conduct an effective defense were lost in the fighting across France.[5]

The First US Army sector was to penetrate the *Westwall* north of Aachen where the German defenses appeared the weakest. The 30th Infantry Division was tasked to make the break in the German defenses then swing to the south to link up with the 1st Infantry Division, under VII Corps, near the town of Würselen, thus encircling the city of Aachen. The 2d Armored Division stood by to exploit the penetration by crossing the Wurm River and push an additional nine miles to seize the crossings of the Roer River. To draw German attention and forces away from this main effort, the 29th Infantry Division was to make limited objective attacks along the corps north flank. The target date for the assault was 1 October. It was hoped the city could be taken "on the bounce" or bypassed so the Allied drive could continue unabated to the Rhein River, but that was not to be. Stiffening German resistance and the salient made by Aachen made its capture necessary.[6]

Major General Leland S. Hobbs, commander of the 30th Infantry Division, chose a narrow one-mile sector nine miles north of Aachen to make the penetration along the Wurm River. This area was chosen to avoid the stronger defenses near Geilenkirchen and the dense urban centers close to Aachen. Hobbs could have selected an assault sector further south to facilitate a quicker juncture with VII Corps, but ruled that option out. He saw that the roads in the north of his area of operations were better suited for supply routes, and he could avoid a number of potential urban fights.

The 1st Infantry Division was positioned to the right of the 30th Infantry Division and it too sought the best routes to reach Aachen. Active patrolling by the 26th Infantry Regiment ascertained the approximate troop strength of the Germans in this sector and assessed the roads and avenues of approach to the city. These reconnoiters confirmed that Aachen was going to be heavily defended.[7] (See Map 2.)

Meanwhile, the Germans had reinforced this sector with the relatively fresh *183d Volksgrenadier Division*. The fortifications of the *Westwall* were manned and reserves assembled from this division and an assortment of scratch units. The Germans were holding the line west of the city of Aachen with seven battalions of about 450 men each. From Geilenkirchen to Rimburg there were two battalions of the *330th Infantry Regiment* of the *183d Volksgrenadier Division*. South of Rimburg was five battalions of the *49th Infantry Division*. There were four artillery battalions firing into the 30th Infantry Division's zone plus a battery of 210mm guns and two large-caliber railroad guns. There were only a few tanks available along this portion of the front. The Germans had at least one battalion from each of the three regiments of the *183d Volksgrenadier Division* situated for quick counterattacks. Positioned west of Aachen was the *116th Panzer Division,* an operational reserve. These German units were not the homogenous units of the past; at this stage of the war, these units were a collection of ad hoc and composite elements and unit cohesion suffered accordingly.[8]

Hobbs ordered his men to prepare for a general attack in this sector to cut off Aachen from the north and for the eventual link up with the 1st Infantry Division attacking from the south. On 26 September, massed

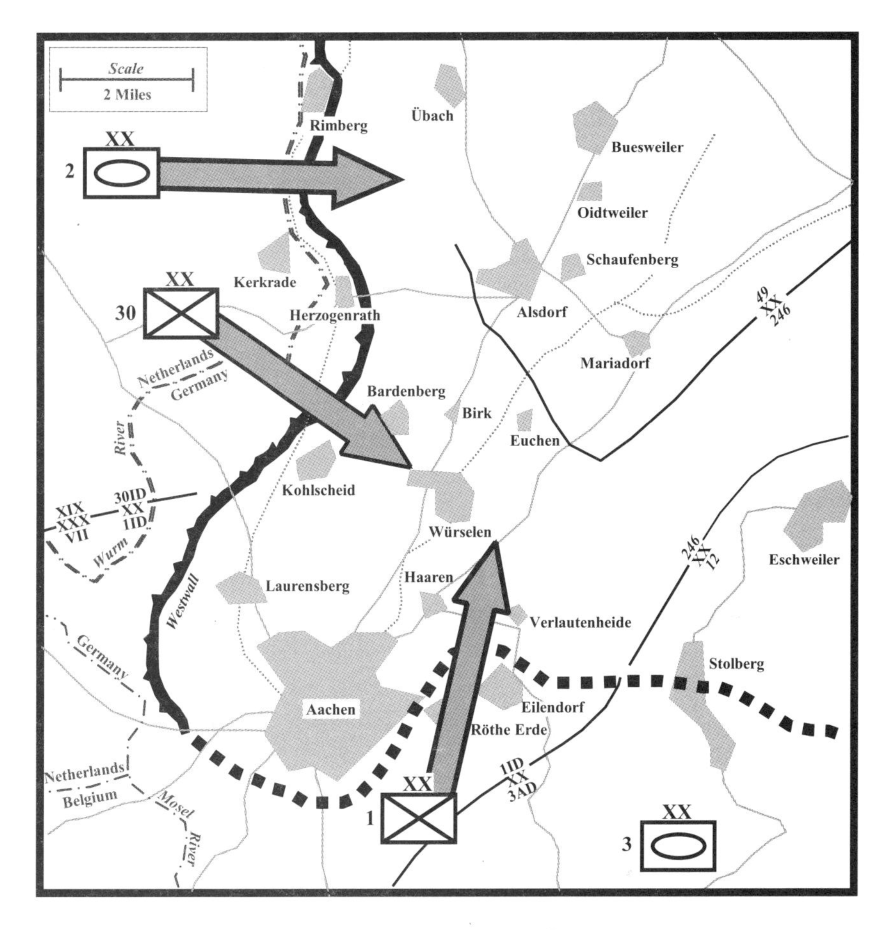

Map 2. The plan to take Aachen, October 1944.

artillery began the systematic attempt to destroy all enemy pillboxes in the 30th Infantry Division's sector. For this effort, 26 battalions were used, including units from the 2d Armored Division, 29th and 30th Infantry Divisions; four battalions attached to the 30th Infantry Division; eight battalions of XIX Corps Artillery; and three battalions from First Army Artillery. After the general bombardment of the front, the artillery was to shift targets a few hours before the infantry assault. At this phase, the artillery was to target enemy antiaircraft guns to protect the planned preliminary air strikes by medium bombers and conduct counterbattery fire. The air barrage was meticulously planned to avoid the fratricide experienced at Normandy.[9]

The regiments of the 30th Infantry Division conducted refresher training in the procedures for assaulting pillboxes and working with armor.

No training was conducted on urban operations, but the skills practiced would prove useful in the weeks to come. Similar preparations were made in the 1st Infantry Division sector directly facing Aachen. There the 18th and 26th Infantry Regiments prepared to assault the German *Westwall* defenses. The 18th Regiment was positioned to drive north from the area around Eilendorf and affect a juncture with the 119th Infantry Regiment north of Haaren, thus sealing off Aachen. The 26th Infantry Division would make the actual assault into the city. Two companies of medium tanks from the 3d Armored Division, numbering 20 vehicles, were allocated to the 26th Infantry Division as a counterattack force.[10]

The assault on the *Westwall* began the morning of 2 October as Allied aircraft streamed from the overcast skies to drop their bomb loads on the German positions opposite the 30th Infantry Division. Meanwhile, more than 400 artillery pieces fired a rolling barrage to support the infantry attack. The riflemen of the 117th and 119th Infantry Regiments surged forward, crossed the narrow Wurm River on duckboards, and found shelter behind the railroad embankment on the opposite bank. Quickly regrouping, the men began knocking out the pillboxes and bunkers to their front one by one using small arms, grenades, flamethrowers, bazookas, and guts. The teamwork and skills practiced in the days before the attack paid off. Each man knew his job and did it. Unfortunately, tanks and tank destroyers were unable to cross the Wurm River due to the soft banks. They would have to wait to cross the river until the treadway bridge was completed that evening. Although casualties were high, the infantrymen slogged it out making slow but steady progress through the layers of German defenses.[11] (See Map 3.)

Due to the success of the diversionary attacks to the north by elements of the 29th Infantry Division, the Germans' reaction to the assault was slow. The true nature and location of the American main effort became known only after several hours and by then the *49th Infantry Division* had no forces available for a counterattack. The *183d Volksgrenadier Division* had only one infantry battalion. An ordered counterattack eventually came in the form of an assault gun battalion supported by an infantry company from the *183d Volksgrenadier Division*. Due to Allied air supremacy, this small task force could not begin movement until after dark. When it did move, it came under highly effective interdiction fires from the massed American artillery. The German counterattack finally made contact about midnight, but was reduced to two assault guns with some infantry support. Heavy fire from the Americans, including bazookas, forced the Germans to retire.[12]

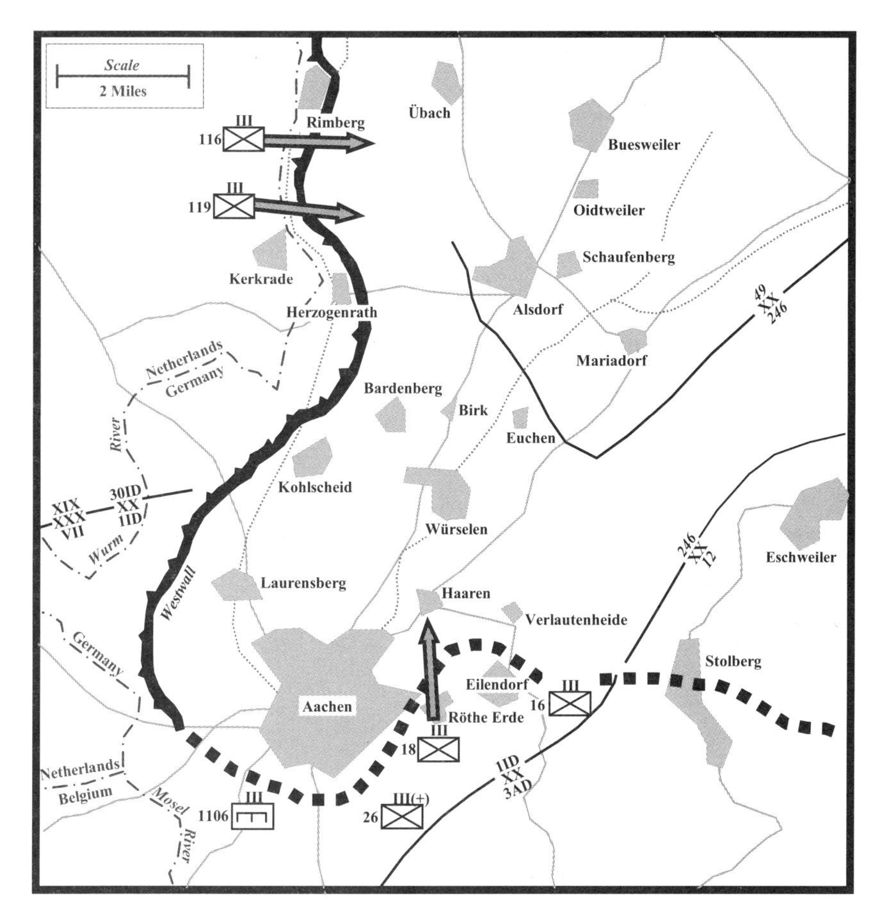

Map 3. Aachen and vicinity, 2 October 1944.

On the morning of 3 October, the tanks and half-tracks of the 2d Armored Division's Combat Command B (CCB) were committed to the small bridgehead formed by the 30th Infantry Division. CCB was to help expand the bridgehead and free the infantrymen for the push to the south and subsequent link up with the 18th Infantry Regiment of the 1st Infantry Division. Intense German shelling and heavy rains throughout the week hindered movement of vehicles in the small pocket, but the American armor pushed across. By nightfall, the infantry and armor had advanced to the northern and western edges of Übach.[13]

The Germans frantically scraped up units to throw at the Americans. To streamline the chain of command, the *49th Infantry* and *183d Volksgrenadier Divisions* were placed under General Wolfgang Lange, commander of the latter. Two assault gun brigades, the *183d Volksgrenadier Division's* organic engineer battalion, two infantry battalions from the *49th Infantry Division*, and an infantry battalion from the *246th Volksgrenadier Division* previously located in Aachen were rushed toward the sector. Delays postponed the counterattack until the morning of 4 October. The heaviest blow from the Germans fell on the 119th Infantry Regiment, but massed artillery fire from the Americans disrupted this and two other attacks in a matter of hours. To the south, the German *27th Infantry Division* launched a strong counterattack supported by eight assault guns and an artillery barrage along the 1st Infantry Division sector. After heavy fighting, the Germans withdrew, unable to push back the soldiers of the Big Red One. Instead, the Americans resumed their advance with the objective of encircling Aachen.[14] (See Map 4.)

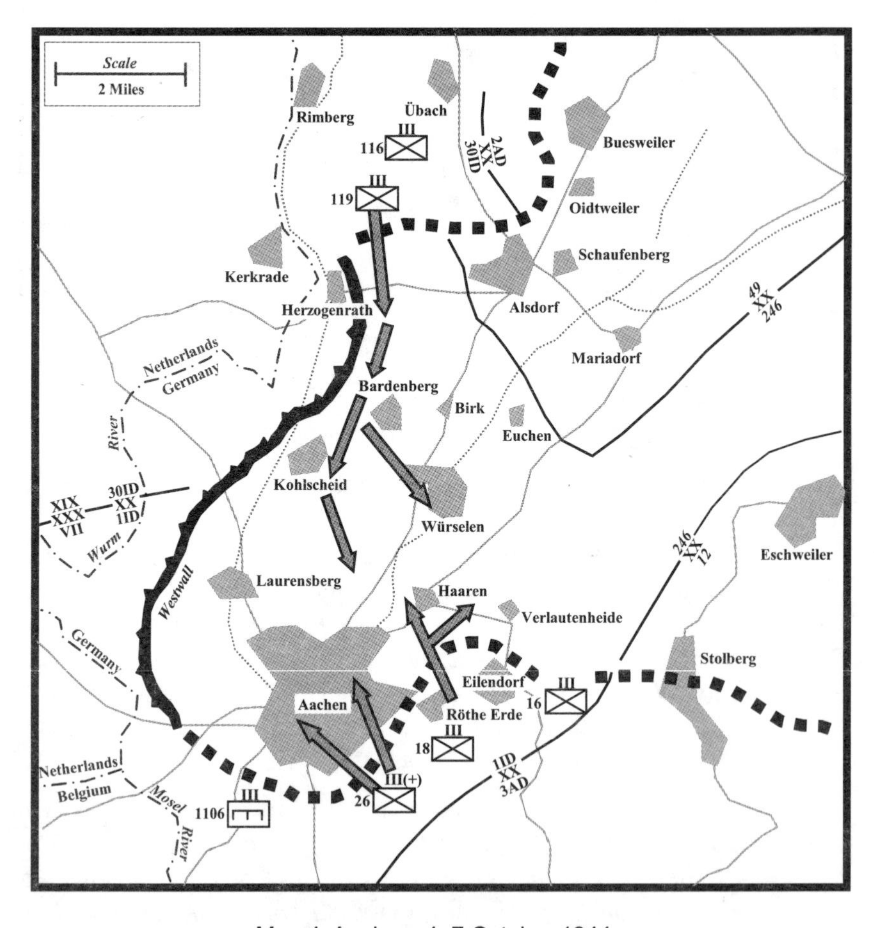

Map 4. Aachen, 4–7 October 1944.

Encircling Aachen

The city of Aachen had a long and proud history dating from Roman times. It was the birthplace of Charlemagne and the capital of the Holy Roman Empire where over 32 emperors and kings were anointed. Prior to heavy Allied bombing, Aachen had 165,000 people, some industry, and coal mining; but, by August 1944 the city had less than 20,000 of its prewar population. As the Allies moved closer, a mandatory evacuation order removed all but about 7,000 of those remaining to safety. Although the city no longer significantly contributed to the war effort, for Hitler Aachen was the center of the First Reich and possessed great symbolic importance. Hitler vowed the city would not be taken, proclaimed it to be a fortress, and expected every soldier to fight to the last. After the war, *Generaloberst* Alfred Jodl, Chief of the *Wehrmacht* Operations Staff, wrote that Aachen did not have any special operational significance, but it was important because it was the first German city attacked. To the troops, the people, and the opponent, its defense to the last was to serve as a shining example of the tenacity with which the Germans would fight for their homeland.[15]

Clearly, the city meant far more to the Germans than to the Americans. The situation the city found itself in was actually an accident of geography, as it was within the successive belts of the *Westwall* and in the path of the advancing American armies. Lying in a valley surrounded by hills, the city offered no strategic or tactical value. The road network was relatively unimportant as the American thrusts north and south of the city revealed adequate roads leading to the Rhein River. Although a rail line passed through the town, it was so badly damaged by bombing that it would take weeks or months to repair. In addition, massive bombing and shelling had reduced Aachen to rubble, leaving very little infrastructure intact. The only tactical value of Aachen to the Allies was its location on the shortest route to the key German industrial region along the Ruhr River.[16]

Colonel Gerhard Wilck, commander of the *246th Volksgrenadier Division,* was the ranking German officer at Aachen. He established his headquarters in the Hotel Quellenhof, a luxurious facility in Farwick Park in the northern portion of the city. Most of the troops present came from the *246th Volksgrenadier Division*, but there were other German troops there. The *34th Fortress-Machine Gun Battalion*, augmented with elements of the *453d Infantry Training Battalion*, was a scratch unit with limited combat value. There were approximately 125 Aachen policemen still in the city and 80 more from Köln (Cologne) that were pressed into the line. Two *Luftwaffe* fortress battalions consisting of a myriad of specialists with no infantry training were also sent to the city. Wilck had no more than five Mark IV tanks with the 75mm high-velocity guns within the city. Within the city perimeter were 19 105mm howitzers of the *76th Motorized Artillery Regiment*, 8 75mm guns, and 6 150mm pieces from the *146th Armored Artillery Regiment*. *Antiaircraft Group Aachen* was formed from a variety of weapons and was deployed mostly as field artillery. As long as communications held, Wilck could also call substantial artillery fire from outside the American ring. The aircraft of the *Luftwaffe* only appeared at night and in small groups, so they were not a major factor in the fight.[17]

As the pincer movement of the 30th and 1st Infantry Divisions encircled Aachen, the 26th Infantry Regiment began its operation to seize the city itself. This available force was numerically inferior to the German defenders. Two battalions were arrayed to assault with one infantry battalion in reserve. Two companies of M4 Shermans and a tank destroyer company were allocated to support the assault.

In the hope to avoid house-to-house fighting, on 10 October Lieutenant General Courtney H. Hodges sent an ultimatum to the German commander giving him 24 hours to surrender or face a massive bombardment. Although some of the few remaining civilians hoisted white flags from their windows, the German military made no offer of surrender. American bombers and artillery pounded Aachen on schedule, beginning at noon on 11 October.[18]

The dire situation was not lost on the Germans, but their ability to respond to it was limited. With the supply routes to Aachen now restricted, it was imperative to launch a strong attack against the 18th and 119th Infantry Regiments blocking the roads. The *3d Panzergrenadier* and *116th Panzer Divisions* were sent to the threatened sector on 10 October, but they were committed piecemeal as the separate elements arrived. The heavy American firepower and some savage close-in fighting repelled these small elements. The Germans continued to send units to relieve Aachen over the next several days, but the thin American line preventing this held. Without reinforcements and resupply, the garrison was isolated and doomed. It would not give up without a fight, and the Americans were preparing to storm the city.[19] (See Map 5.)

Colonel John F.R. Seitz, commanding the 26th Infantry Regiment, prepared his 2,000-man outfit to take Aachen. For the past two days his men had inched into Aachen's eastern suburb of Röthe Erde, getting into position for the main assault in the center of the city. Seitz placed a provisional company along the left of his line to maintain contact with the 1106th Engineer Group, which was holding a defensive line to the west of the city. The 2d Infantry Battalion, under Lieutenant Colonel Derrill M. Daniel,

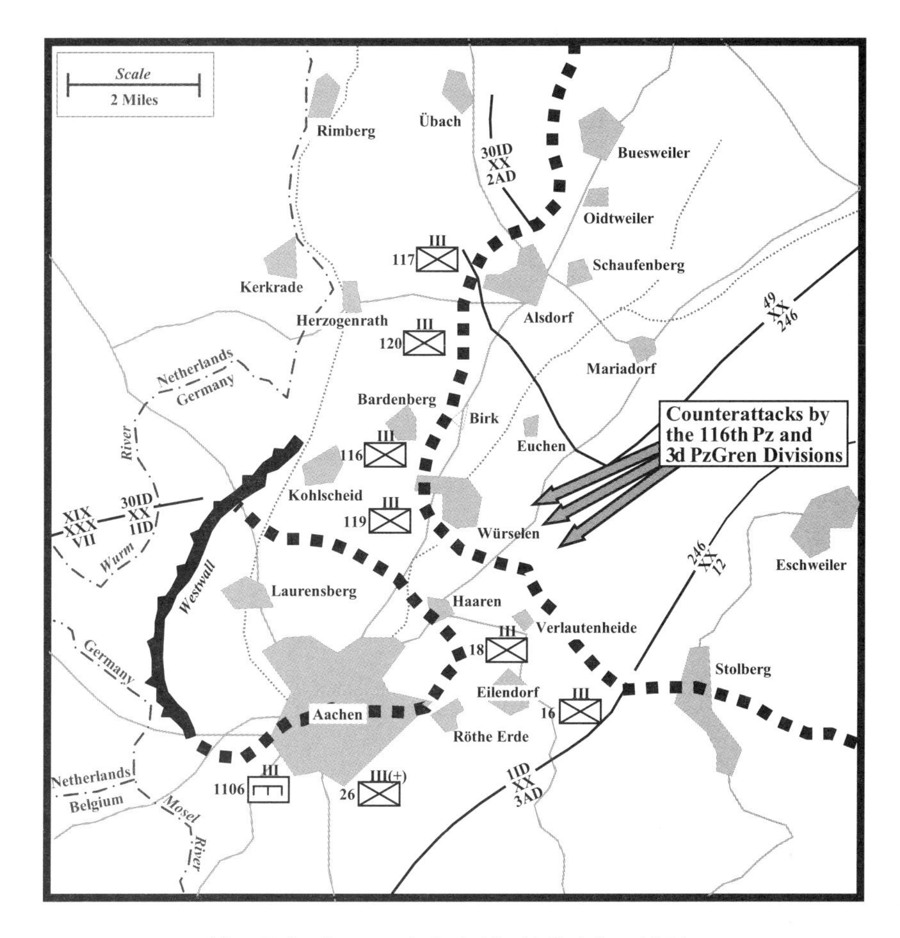

Map 5. Aachen encircled, 10–11 October 1944.

was positioned to press up the Aachen-Köln (Cologne) railroad tracks and into the heart of the city with its narrow medieval streets and masonry buildings. The main effort was the 3d Infantry Battalion, under Lieutenant Colonel John T. Corley. It was to advance northwestward against the modern section of Aachen with its wide streets and factories, then swing northeast to seize three hills that dominated Aachen from the north. These hills were a big public park known collectively as the Lousberg to the Germans and Observatory Hill to the Americans, as there was an observation tower on the crest of the tallest summit.[20] (See Map 6.)

Soon after the surrender deadline on 11 October, approximately 300 Allied planes began their bombing runs under perfect weather conditions. Primary targets were selected along the perimeter of the city by the infantry and marked with smoke by the artillery and mortars. Twelve battalions of artillery joined in the bombardment firing over 5,000 rounds. Although

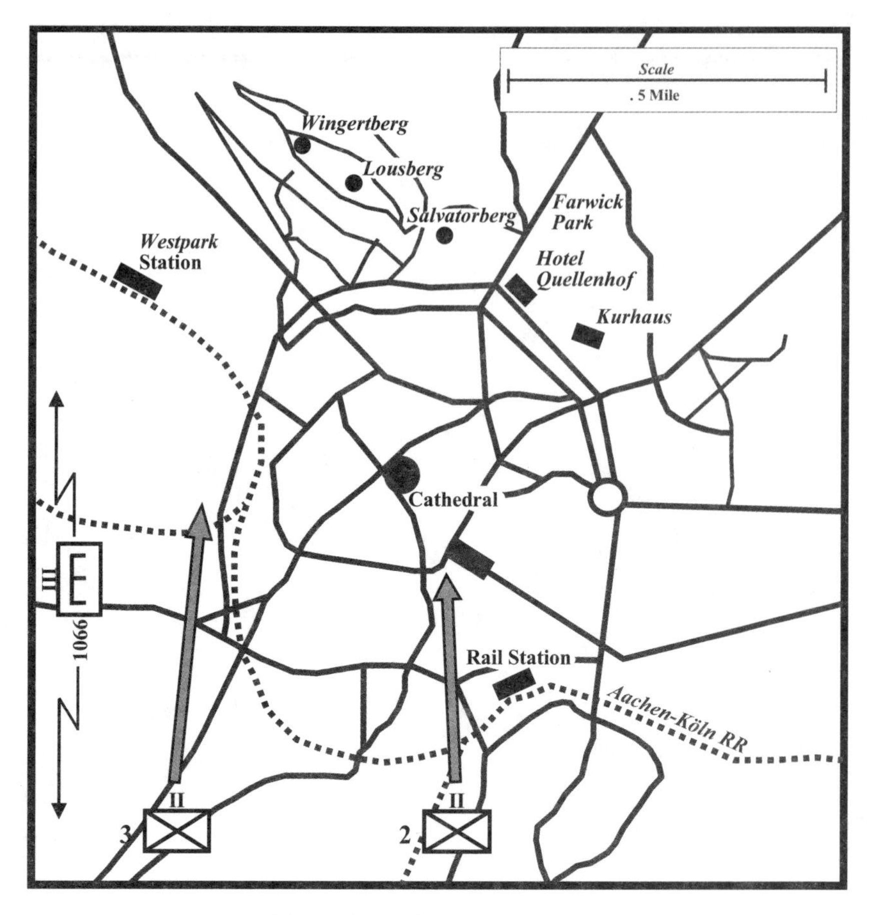

Map 6. Fight for the city center.

the air and ground spotters judged the bombing and shelling as accurate, the infantrymen saw no appreciable lessening of enemy fire.[21]

At 1100 on the morning of 12 October, the 26th Infantry Regiment began its attack on Aachen proper. Daniel had divided his 2d Battalion into small assault teams, and each infantry platoon had an attached tank or tank destroyer. The plan was for the armor to provide direct fire support by keeping a building under fire until the riflemen made the assault. As the assault was made, the tanks would shift fire to the next building or any source of enemy fire. Supported by the fire of each battalion's light and heavy machine guns, the infantry would then enter and clear each room with small arms and grenades. If a particularly tenacious defense was encountered, the Americans were prepared to use bazookas, demolitions, and flamethrowers that were attached to each company headquarters, and two 57mm antitank guns attached from the regiment. The light artillery and mortars would lay a barrage a street or two in front of the advancing infantry, while the heavy artillery and aircraft would pound German positions further in the rear. To choreograph all of these elements, Daniel designated a series of checkpoints based on street intersections and prominent buildings. No unit was to advance beyond a checkpoint until it established contact with an adjacent unit. Additionally, units were assigned specific zones of advance down to platoon level. Units were to halt each night to prevent fratricide and to rest, resupply, and consolidate. To facilitate the movement of logistics and evacuate wounded, Daniel improvised a supply ammunition train using the tracked M29 cargo carriers known as Weasels.[22]

From a doctrinal standpoint, the organization for the attack was conventional as tanks and artillery were habitually attached to infantry formations for close fire support. The tactics employed in the use of armor was not much different than the fighting in the hedgerows during the previous summer. Tanks provided the heavy firepower while the infantry protected the armored vehicles from lurking German infantry armed with antitank weapons. In the drive across France, the riflemen and armor crewmen had fought together and all of these veteran outfits knew the capabilities and limitations of the other.

The Fight for Aachen

The 2d Battalion had not advanced to the first checkpoint before the plan temporarily broke down. Within a matter of minutes of entering the outskirts of Aachen, over 20 Americans lay on the cobblestones shot in the back. Before their startled comrades could react, their attackers had disappeared the way they had come, scuttling into the cover of the sewers. These sewers posed a special problem, necessitating the locating and sealing of all manhole covers. The American infantrymen also learned that each house had to be thoroughly cleared of both enemy troops and civilians before passing on. Speed became less important than thoroughly checking each building for hidden Germans.[23]

The 3d Battalion launched its attack toward Observatory Hill/Lousberg but found the route blocked by heavily defended apartment buildings. Fighting became measured by the gains in buildings, floors, and rooms. *Panzerfausts* quickly knocked out two supporting Sherman tanks, but one was recovered. It was apparent that some of the buildings in this sector were impervious to tank fire. To break the tenacious defense, Corley called up a self-propelled 155mm artillery piece and employed it in the same direct support role as the tanks. The big gun proved its worth as it practically leveled a sturdy building with one shot. Although it was unusual to use artillery in a direct-fire mode in a city, Seitz sent another big gun to support the 2d Battalion. Results trumped doctrine. By nightfall, the 3d Battalion had reached the base of the hills, but in some points in this sector the infantrymen were just holding on to their positions under heavy German fire.[24]

The fighting in Aachen settled into a routine for the Americans who learned the fine points of urban warfare along the way. Typically, a tank fired on the building just ahead of its supporting rifle platoon, suppressing enemy fire until the soldiers could enter and clear the structure with hand grenades and automatic fire. The Americans were rightfully concerned about the ever-present threat of *panzerfausts*. Infantrymen covered the vulnerable tanks, which in turn covered the infantry. In practice, the tanks and tank destroyers usually stayed one street back from the advancing infantry, creeping forward or around a corner to engage their targets. Once the block was cleared, the armored vehicles would dash forward to the newly cleared street. There was no attempt to avoid collateral damage, which was immense.[25]

One aspect of the fighting was unconventional. For several days prior to the attack on Aachen, a detachment of US Army Rangers was operating under the direction of the Office of Strategic Services (OSS). Donning German uniforms with the correct papers and accoutrements, the Rangers were organized into small teams and were to penetrate enemy lines to conduct sabotage and raids. Every man spoke German fluently and each had spent months training for such operations. Working under the cover of darkness, one was successful in entering Aachen and succeeded in destroying a signals center. The Rangers then positioned two machine guns covering the barracks of a German quick-reaction force. Using signal flares and pre-arranged American artillery to rouse the enemy, the Rangers cut the Germans down as they emerged from the bunker.[26]

On 13 October, the 2d and 3d Battalions of the 26th Infantry Regiment continued the attack, making slow but steady progress. During the fighting in the industrial section, a German 20mm antiaircraft gun dispersed the infantry covering two M4 Shermans. A *panzerfaust* destroyed one tank and damaged another. Three very brave infantrymen removed the dead and wounded and miraculously drove the tank back to safety. A juncture was formed between the two battalions within the city, but there was plenty of fighting still to come.[27]

Wilck and his men were making a determined resistance, but the German commander was alarmed at the American's success in advancing toward Observatory Hill/Lousberg. He had expected the main American attack to come directly into the city from the south and deployed his forces accordingly. The enemy attack on the hills overlooking the city faced the weaker defenses on the west side of Aachen. Wilck immediately deployed about 150 men from his own *404th Infantry Regiment* to the threatened sector. Unable to send more troops to that endangered position and believing his headquarters was about to be overrun, the German commander appealed for more reinforcements. The small *SS-Battalion Rink* from *Kampfgruppe Diefenthal* was holding a section of the *Westwall* to the north and was busy containing one of the 30th Infantry Division's supporting attacks. By nightfall on 14 October, this *SS Battalion* was disengaged from the fighting to the north, reinforced with eight assault guns, and sent toward the city. It had lost nearly 50 percent of its personnel strength in recent fighting and would not have time to reorganize or replenish. This combined force would not arrive for 24 hours.[28]

Corley's 3d Battalion renewed its attack on 15 October with the assistance of close support from 4.2-inch chemical mortars. By midday the infantry had taken a number of key buildings, but the walls of the Hotel Quellenhof proved resilient. Corley was bringing up the 155mm gun and a reserve company to deal with this obstacle when the Germans launched a battalion-size counterattack. The American infantry held for about an hour, but the Germans, supported by assault guns, applied the same tactics as used against them. The men of the *404th Regiment* and *SS-Battalion Rink* forced one American company to withdraw then swept southward to hit the next in line. The Germans were successful in knocking out one tank destroyer, one antitank gun, and one heavy machine gun. The Americans held though and by 1700 the German attack was spent. Due to this attack and the threatening German movements to break the ring around Aachen, Seitz was ordered to hold the 26th Infantry Regiment in place to see how the situation developed. The time was well used to consolidate, resupply, and reorganize.[29]

On 16 October, the 2d Battalion used the lull to take care of a large pillbox sighted in its area. Daniel brought up his 155mm gun using tank destroyers to cover its movement and infantry to cover the tank destroyers. One of the tank destroyers knocked holes in the foot of a building to create a clear field of fire for the big gun. Once in position, the heavy155mm gun made quick work of the enemy pillbox, which turned out to be a camouflaged panzer. In addition, the 1106th Engineer Combat Group, previously holding a line of containment to the west of Aachen, moved forward to gain contact with the 26th Infantry Regiment. This freed the infantry for the continued advance.[30]

The Germans clearly recognized that unless the *3d Panzergrenadier* and the *116th Panzer Divisions* could break the ring around Aachen, the city was lost. Despite the German resistance, the American 119th Infantry Regiment reached the juncture point with the 18th Infantry Regiment on 16 October. The 18th Infantry Regiment had been delayed by the German attacks in the 1st Infantry Division's sector, but was able to push some three miles to Haaren and completely cut Aachen off. Three days of bitter fighting saw these units strike the Americans to the east of the city, but the attacks were poorly coordinated and no breakthrough occurred. Attempts by the Aachen garrison to support these efforts from inside the pocket were unsuccessful. Continuing to hold off fierce German attacks sealed the fate of the city. By the evening of 19 October, the senior German commanders in the sector decided to cease all efforts to relieve Aachen and prepare to meet the next Allied blow. Wilck was faced with fighting to the last man or surrender.[31]

The Final Push

Once the threat of a German relief of Aachen was eliminated, General Joseph L. Collins, commander of VII (US) Corps, decided to employ a large force of armor to decide the issue quickly. One tank battalion and one armored infantry battalion from the 3d Armored Division were organized as Task Force Hogan and sent to join Corely's 3d Battalion on its assault to take Observatory Hill/Lousberg. Once the hill was taken, Task Force Hogan was to proceed to the village of Laurensberg, two miles northwest of Aachen and a point in the *Westwall* still held by the Germans. Additionally, the 2d Battalion, 110th Infantry Regiment from the 28th Infantry Division was attached to the 26th Infantry Regiment to be used in a defensive role, occupying buildings as the assault troops pushed forward.[32] (See Map 7.)

The assault was renewed on 18 October. The 3d Battalion of the 26th Infantry Regiment pushed forward, recapturing the ground lost three days earlier and seizing Hotel Quellenhof. The 2d Battalion continued its methodical advance into the heart of Aachen. Task Force Hogan maneuvered into position to attack and take Observatory Hill/Lousberg. This was done with great difficulty due to the soft ground and German resistance. The next day the Americans advanced relentlessly. The armor of Task Force Hogan pushed toward Laurensberg, but was diverted to the east because the 30th Infantry Division got there first. Daniel's 2d Battalion seized the main railroad station in the heart of Aachen and pushed north. German resistance was crumbling.[33]

On 19 October, Wilck issued the order for the day that said, in part,

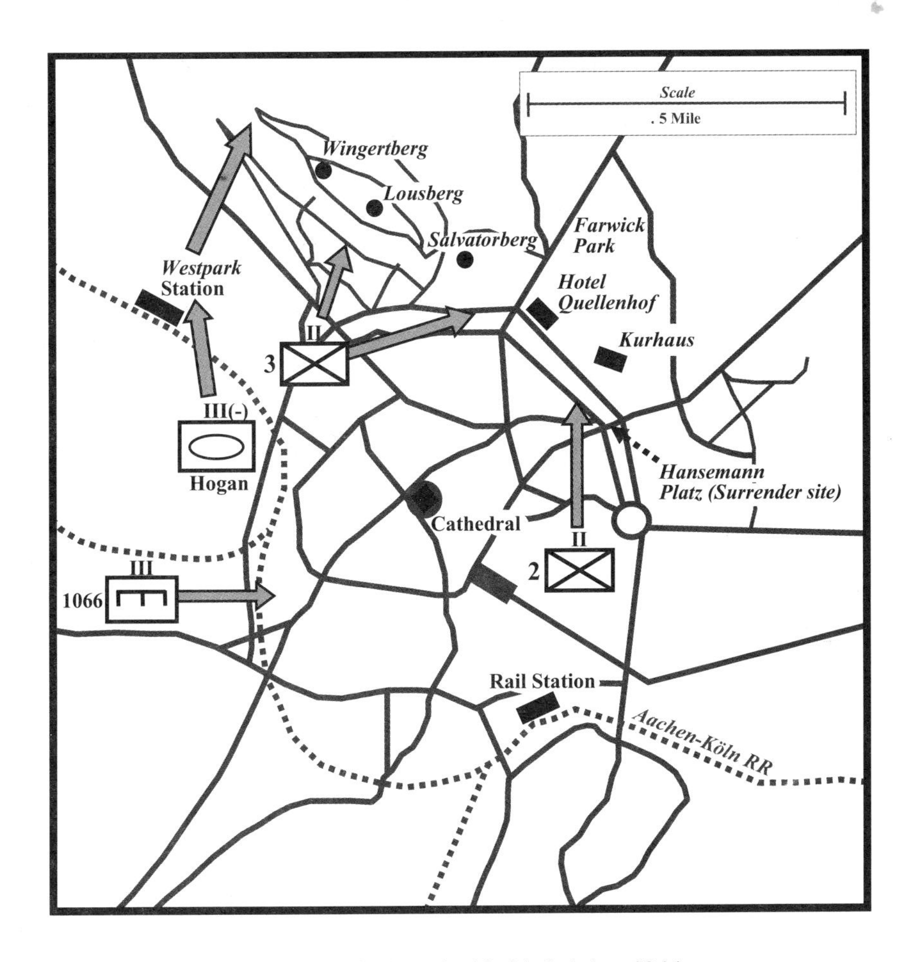

Map 7. The final push, 18–21 October 1944.

"The defenders of Aachen will prepare for their last battle. Constricted to the smallest possible space, we shall fight to the last man, the last shell, the last bullet in accordance with the Führer's orders." This was just part of a stream of such messages sent via radio out of Aachen, but it had no effect on the reality of the hopeless situation. In fact, Wilck had warned his corps commander that he expected the city to fall in the next day or so and asked for a decision for a breakout attempt. Permission was denied. Instead, both sides continued to fight, but hour by hour, Aachen was being lost to the Americans. Wilck was down to about 1,200 effective men and one assault gun. Although artillery spotters were acquiring targets, there was no ammunition for the guns. Every cellar near the Hotel Quellenhof was packed with wounded, and medical supplies were nearly gone. So was the will to fight.[34]

The end came rather suddenly after noon on 21 October as Corley's 3d Battalion prepared to engage what they thought was an air raid shelter with one of the 155mm self-propelled guns. It was in fact Wilck's headquarters, and he was finally ready to surrender. Using two captured Americans to get their comrades' attention to cease firing, the Germans emerged from the structure and surrendered. The rest of the German garrison laid down their arms when they received word of the surrender. They and the civilians were evacuated and the war moved on to the east. Quite interesting is that no subversive group rose up to challenge the American occupation. The Germans had not prepared for such action and, in any case, the evacuation of all civilian personnel from the city made it impossible.[35]

American casualties in the final seven days of street fighting totaled 75 killed, 414 wounded, and 9 missing in action. The Germans lost close to 2,000 men killed or wounded and over 3,400 prisoners within the city. The 30th Infantry Division captured over 6,000 prisoners in the fighting in and around Aachen.[36]

In Retrospect

Overall, the American effort to reduce and take Aachen went very well, but it took aggressive fighting and a high degree of adaptation and flexibility. The soldiers and armored vehicle crewmen were not specifically trained for city fighting, but were able to apply the principles of firepower and teamwork to full effect. Innovative use of the 155mm self-propelled guns and tanks to reduce enemy positions is an example. Another example is Daniel's tactical control measures to ensure the destruction of each German strongpoint and prevent fratricide. But the battle for the city itself was quite conventional, even linear, when analyzed. The tactics employed emphasized fire and maneuver, the standard for the day, and closely resembled the procedures used in the hedgerows and through the numerous towns and villages in the drive across France. The tanks and tank destroyers were used as mobile platforms to bring heavy ordnance to where it was needed by the infantrymen. Armor was in the supporting role, and not engaged in bold offensives leaving the infantry to catch up. The riflemen bore the brunt of the fight for Aachen, but counted on the armor to make their job possible. This operation just happened to be at a much larger scale.[37]

By this time the German army had experience in city fighting, but this experience did not necessarily transfer to the men fighting in Aachen. These men were from a relatively ad hoc unit, namely the *246th Volksgrenadier Division* that was formed only a few weeks earlier. The *SS-Battalion Rink* was a veteran outfit, but was comparatively small in number. To their detriment, the Germans never massed a force large enough to break through to Aachen. With two exceptions, the counterattacks against the encircling ring of Americans were never more than two battalions and inside the city the largest was by a small battalion-size unit. These attacks were poorly coordinated and usually conducted piecemeal. Conspicuous too was the limited number of German mines and booby traps. Also, the location of Aachen ultimately doomed the city. Nestled in a bowl with overlooking hills, the town was actually indefensible against a strong and determined enemy.

Although the units involved in the fighting were predominately infantry formations, both sides failed to realize much success unless tank support was on hand. In spite of their technical shortcomings, the M4 Shermans and the M10 tank destroyers provided the vital heavy fire support needed to blast through the thick walls shielding the defenders. Although at a serious disadvantage against German panzers and *panzerfausts*, the armor of the American vehicles did offer sufficient protection against small arms fire. When sufficiently protected by the infantry, American armor proved very capable in urban fighting.

The battle for Aachen influenced Army doctrine for years to come. Commanders were warned to avoid committing forces to the attack in urban areas. Nevertheless, seriously outnumbered at Aachen, the Americans were able to take the city after only nine days, and three of those were used to rest and reorganize. In comparison to the fighting during the Battle of the Bulge and the Hürtgen Forest, the casualties suffered by the Americans in Aachen were very light. If anything, the battle for Aachen showed the Americans' ability to adapt and continue to fight. In spite of being outnumbered and fighting in the streets of the enemy's home city, the Americans won the day.

Notes

1. Bryan Perrett, *Iron Fist: Classic Armoured Warfare Case Studies* (London: Arms and Armour Press, 1995), 121.

2. Christopher R. Gabel, "Knock 'em All Down: The Reduction of Aachen, October 1944," in *Block by Block: The Challenges of Urban Operations,* ed. William G. Robertson (Fort Leavenworth, KS: U.S. Army Command and General Staff College Press, 2003), 72–73. Dr. Gabel also wrote an excellent account of the battle of Aachen titled "Military Operations on Urbanized Terrain: The 2d Battalion, 26th Infantry at Aachen, October 1944," in *Combined Arms in Battle Since 1939,* ed. Roger J. Spiller (Fort Leavenworth, KS: U.S. Army Command and General Staff College Press, 1992).

3. Belton Y. Cooper, *Death Traps: The Survival of an American Armored Division in World War II* (Novato, CA: Presidio Press, 1998), 19–22. Cooper's book describes the shortcomings of American armor design. He was commander of a tank recovery unit and saw the damage firsthand. Incidentally, the M26 Pershing tank was equal to the German panzers, but did not see action until very late in the war and in very few numbers. See also Ian V. Hogg, *Armour in Conflict: The Design and Tactics of Armored Fighting Vehicles* (London: Jane's Publishing Company, 1980), 165–167. Refer to R.P. Hunnicutt, *Sherman: History of the American Medium Battle Tank* (Novato, CA: Presidio Press, 1978) for a detailed history of the M4 Sherman.

4. Gabel, "Knock 'em All Down," 65. See also Roman J. Jarymowycz, *Tank Tactics: From Normandy to Lorraine* (London: Lynne Rienner Publishers, 2001), 263.

5. Ibid., l, 66. See also H.R. Knickerbocker, *Danger Forward: The Story of the First Division in World War II* (Washington, DC: Society of the First Division, 1947), 279–280.

6. Charles Whiting, *Bloody Aachen* (New York: PEI Books, Inc., 1976), 29–30. Elbridge Colby, *The First Army in Europe,* US Senate Document 91–25 (Washington, DC: US Government Printing Office, 1969), 105–106.

7. Knickerbocker, 261.

8. Heinz G. Guderian, *From Normandy to the Ruhr: With the 116th Panzer Division in World War II* (Beford, PA: The Aberjona Press, 2001), 130.

9. Charles B. MacDonald, *US Army in WWI: The Siegfried Line Campaign* (Washington, DC: US Government Printing Office, 1984), 253–254. Friendly bombs hit the 30th Infantry Division twice during the Normandy campaign. Hobbs wanted the bomb runs made on a parallel approach to avoid repeating the mistakes of the past, but was overruled. A total of 360 medium bombers and 72 fighter-bombers were allocated to the Aachen operation.

10. Knickerbocker, 261.

11. Irving Werstein, *The Battle of Aachen* (New York: Thomas Y. Crowell Company, 1962), 34–36.

12. Guderian, 208.

13. Werstein, 106.

14. MacDonald, 276–279. Guderian, 209–210.

15. MacDonald, 307–308. Gabel, "Knock 'em All Down," 66–67, 82. Guderian, 133.

16. Knickerbocker, 283. Gabel, "Knock 'em All Down," 63.

17. Guderian, 130. MaDonald, 308. Knickerbocker, 260. See also Gabel, "Knock 'em All Down," 71. The *246th Volksgrenadier Division* at Aachen had three of its seven authorized battalions and was made up largely of survivors from other divisions shattered in battle. Command of Aachen passed through at least two men prior to the battle. For brevity and clarity, Wilck is named here as the commander. He assumed this unlucky position on 12 October.

18. Knickerbocker, 262. MacDonald, 307. Werstein, 114–116. Hodges' ultimatum was brusque and simple: surrender or the city would be destroyed. This message was delivered through the lines and by leaflet drops on Aachen.

19. Knickerbocker, 262.

20. Gabel, "Knock 'em All Down," 67, 73. Whiting, 110–111.

21. MacDonald, 309. Gabel, "Knock 'em All Down," 68.

22. MacDonald, 310. Knickerbocker, 263. Gabel, "Knock 'em All Down," 75. Whiting, 136–138.

23. Whiting, 138.

24. MacDonald, 312. See also Gabel, "Knock 'em All Down," 77. Whiting, 138–139. The 155mm gun, called the Long Tom, fired a massive 95-pound armor-piercing projectile at a muzzle velocity of 2,800 feet per second.

25. Gabel, "Knock 'em All Down," 77.

26. Whiting, 143–146. This attack knocked out most but not all German communications within Aachen. Wilck still had a land-line system and radio communication to the outside.

27. Knickerbocker, 263. Gabel, "Knock 'em All Down," 80.

28. Knickerbocker, 263. Guderian, 218. Whiting, 115. *SS-Battalion Rink* was detached from the *1st SS Liebstandarte Division*, which was preparing for the Ardennes offensive. There was friction between Colonel Wilck and Sturmbahnführer (Major) Herbert Rink, as the latter was reluctant to take orders from a *Heer* (Army) officer. As a member of the *Waffen-SS* and veteran of years in Russia, Rink felt his orders should come through the *SS* chain of command.

29. Gabel, "Knock 'em All Down," 80.

30. MacDonald, 313. Gabel, "Knock 'em All Down," 78. A 155mm gun was later used against a sniper position located in a church steeple.

31. MacDonald, 314, Guderian 217, 220–221. *SS-Battalion Rink* was initially successful in pushing back the Americans in the vicinity of the Hotel Quellenhof, but heavy mortar and artillery fire finally repulsed the attack.

32. Knickerbocker, 264.

33. Ibid., 264. Colby, 107.

34. Guderian, 225–226. MacDonald, 315. Whiting, 155, 177–178. Rink ordered the surviving 40 men of his battalion to break up into small groups and

attempt to infiltrate their way through the American lines. Few made it, but Rink did. He was promoted to *Obersturmbahnführer* (lieutenant colonel) and fought in the Ardennes a few weeks later. He survived the war.

35. MacDonald, 317. Wilck had sent two officers out of the bunker under a white flag, but both were shot by the Americans as they emerged. Staff Sergeant Ewart Padgett and Private First Class James Haswell, both from the 1106th Engineers, volunteered for the job. Wilck survived the war. The *246th Volksgrenadier Division* was reconstituted and finished the war in Poland and Czechoslovakia. See also Gabel, "Knock 'em All Down," 82–83.

36. MacDonald, 317. Knickerbocker, 265.

37. Gabel, "Knock 'em All Down," 84.

Chapter 2

Pattons to the Rescue: Hue, Vietnam, 1968

The climatic Tet Offensive of the Vietnam War began on 31 January 1968. One of the most bitter and hard-fought battles occurred at the ancient city of Hue, lasting four weeks and costing 142 American lives. In the battle for Hue, the US Marines of the 1st and 5th Regiments fought alongside the 1st Infantry Division of the Army of the Republic of Vietnam (ARVN) and were supported by elements of the US Army 1st Cavalry Division (Airmobile). The armored forces employed in this fight provided relevant insights for the use of tanks in urban terrain.

The deployment of tanks to Vietnam was a bit of an accident. Troop units were sent to Vietnam incrementally over a period of years and without a long-range strategy. Military planners often did not examine the table of organization and equipment of these units and were sometimes surprised to see what equipment arrived in theater. When US Marine battalions were sent into country to guard airfields, the planners did not realize they had tanks as part of their organizational structure. Once in country the armored forces of the Marines, and the US Army for that matter, did not have doctrine established for using tanks in counterinsurgency operations or for combat in large cities. The doctrine for armor at the time envisioned open country fighting as in a war with the Soviet Union on the European plains. Marine armor doctrine was principled in the use of tanks for direct fire support of the riflemen, and virtually no urban training of any kind was conducted during the Vietnam era by any service.

On their arrival in Vietnam, the US military forces adapted to the situation and used units like the US Army's 11th Armored Cavalry Regiment in search and sweep missions, where their mobility, protection, and heavy firepower were decisive. However, road and convoy security operations were the norm for the numerous tank units dispersed throughout the country. The South Vietnamese saw the value of armor and formed their own cavalry regiments.[1] (See Map 8 for an overview of the I Corps zone of operations in Vietnam.)

The main battle tank of the Americans during the Vietnam War was the M48A3 Patton tank with a 90mm rifled gun. This tank was developed during the early 1950s to counter the Soviet threat in Europe. As an open-country tank, the 52-ton M48 had an advanced fire control system consisting of a stereoscope range finder, range indicator, and an early fire control computer system designed to engage targets at long range. In Vietnam, the M48 proved capable of operating in most terrains, and it

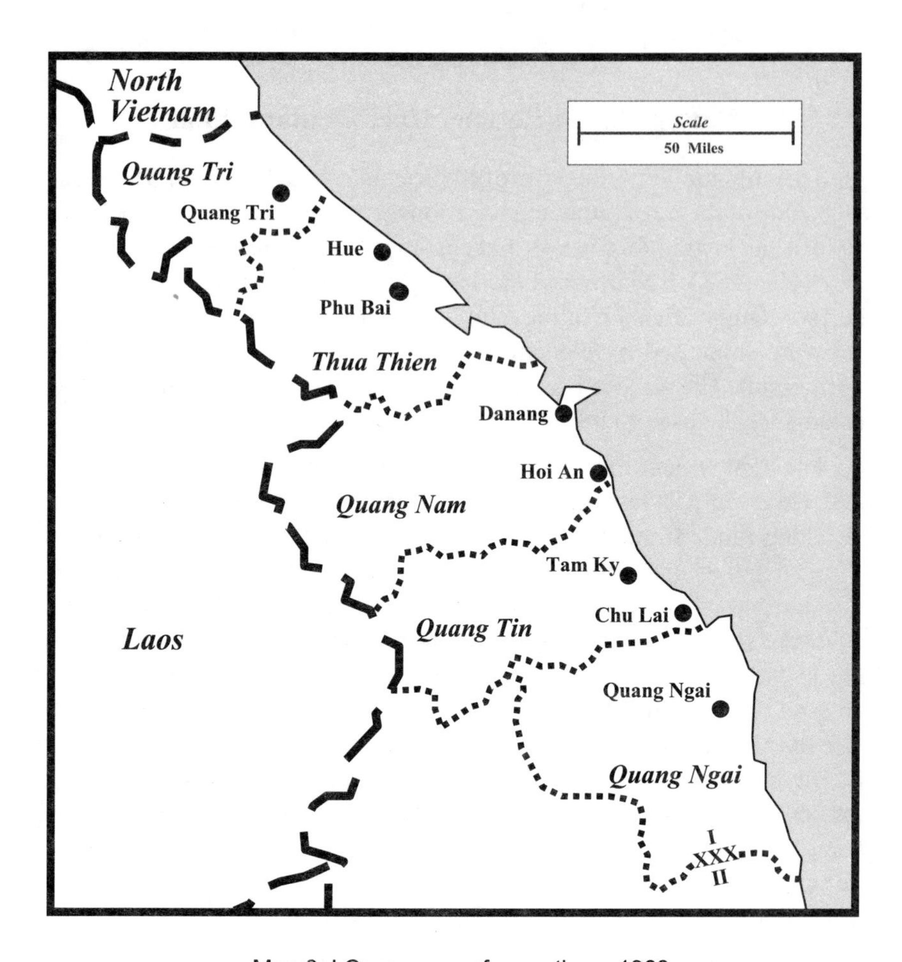

Map 8. I Corps zone of operations, 1968.

was common practice to move the loader from his station in the turret to the back deck with a machine gun for close-in protection. As there were few actual tank-on-tank battles, the M48 provided adequate shelter for its crew from small arms, mines, and rocket propelled grenades, such as the commonly encountered Soviet-made B-40. Few antitank rounds were carried, so most main gun ammunition consisted of high explosive, white phosphorous, and flechette (beehive). A weakness of the M48 was its limited night-fighting ability. The M48 had to rely on illumination rounds fired by artillery and mortars or using the large xenon light mounted above the main gun. The latter method brilliantly illuminated a target, but also broadcasted the location of the tank to the enemy.[2]

In 1968, Hue was the third largest city in South Vietnam with more than 160,000 residents. Located about 60 miles south of the demilitarized zone and six miles west of the coast, the Perfume River bisects Hue, with the modern section on the south bank and the ancient district to the north. The old walled portion of the city, called the Citadel, was about three square kilometers of stone buildings and narrow streets, surrounded on three sides by the river. Built by Emperor Gia Linh in the early 19th century, the Citadel contained the Imperial Palace with its formal gardens and parks, private residences, and market places. A moat encircled most of the old city reinforced by two massive stone walls. South of the river was the modern city consisting of the university, stadium, government administration buildings, hospital, provincial prison, and a few radio stations. Hue was revered as the cultural center of the country, and all sides refrained from combat operations there. As a result, Hue became rather detached and aloof from the war, aside from the Buddhist uprisings in 1963 and 1966. This complacency came to a sudden and bitter end.[3] (See Map 9.)

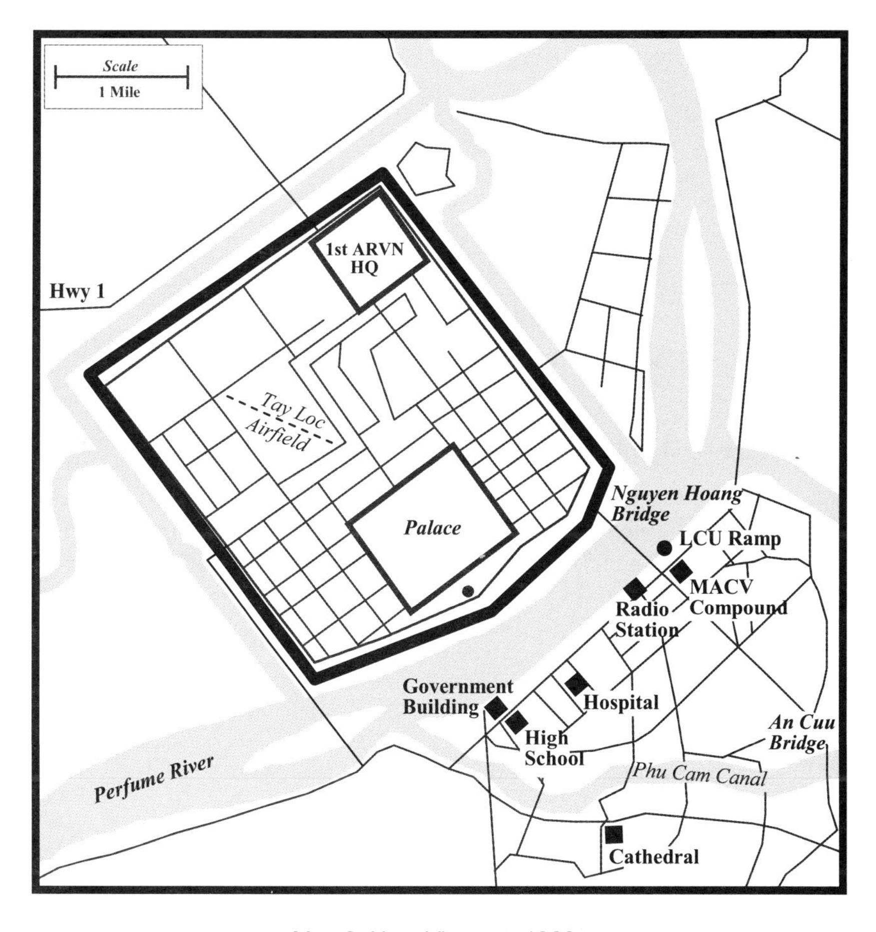

Map 9. Hue, Vietnam, 1968.

The Vietnamese in Hue and the surrounding countryside planned the traditional celebration of Tet in late January 1968. With the traditional 36-hour Tet cease-fire in place on 29 January, many ARVN soldiers and most government officials were on leave or off duty and enjoying the holiday season with their families. Alerted by numerous indicators of enemy activity and truce violations, the US Military Assistance Command Vietnam (MACV) and the joint General Staff of the Republic of Vietnam officially terminated the cease-fire on 30 January. In the I Corps sector, Brigadier General Ngo Quang Truong, the commanding general of the 1st ARVN Infantry Division headquartered in Hue, initiated a heightened security posture and placed his units in a state of increased readiness. Truong canceled leaves; however, most of the troops were already gone and unable to rejoin their units. The only ARVN forces in the city itself were the division staff, the division headquarters company, the reconnaissance company, and a few support units. Additionally, there was Truong's personal guard, the elite Black Panther Company, which he had positioned on the Tay Loc Airfield in the Citadel.[4]

The Attack on Hue

Truong's precautions dampened the holiday mood in Hue, but they were not without merit. The *Viet Cong* was indeed preparing for a major offensive under cover of the Tet cease-fire. On the night of 30 January, *Viet Cong* forces heavily supported by *North Vietnamese Army (NVA)* units launched a massive assault throughout South Vietnam. They attacked all major cities, and although most assaults were repulsed, the enemy forces seized the American embassy in Saigon and the city of Hue.

The North Vietnamese communists had planned to take and hold Hue and use it as a rallying point for their sympathizers in the south. The *Viet Cong* had supplied the *NVA* planners with information on the defenses of the city, the deployment and routine of military and police patrols, and the identities and activities of political opponents, government officials, and foreigners. The *NVA* divided the city into four tactical areas, each with its own priority targets, special missions, and local political leaders. *NVA* forces received an activity schedule for the first three days of occupation and specific instructions for handling prisoners in different categories. Important prisoners were to be evacuated as soon as possible, but they and others were to be executed if evacuation proved impossible.[5]

The *4th* and the *6th NVA Regiments* conducted the main attacks on Hue. The three primary objectives of the *6th NVA Regiment* were the Mang Ca headquarters compound, the Tay Loc Airfield, and the Imperial Palace, all located in the Citadel. The *4th NVA Regiment's* objective was the modern sector south of the Perfume River, including the provincial capital building, the prison, and the MACV advisors compound. Additional objectives for this regiment included radio stations, the police station, the National Imperial Museum, homes of government officials, and recruiting offices. To achieve surprise, a number of assault teams and sappers infiltrated Hue disguised as farmers, with their weapons and uniforms hidden in baggage and under civilian clothing. The *Viet Cong* and *NVA* personnel were able to infiltrate the holiday crowds in the hours before the festivities and take up their predesignated positions. At 0233 on 31 January, a four-man North Vietnamese sapper team dressed in ARVN uniforms killed the guards and opened the west gate to the Citadel. The *6th NVA* then entered the old city en masse. In similar scenes throughout the Citadel, the North Vietnamese regulars and *Viet Cong* poured into both halves of Hue using rockets and mortars in support. The surprise was complete. The enemy troops quickly captured all but a few buildings of the modern city on the south bank of the Perfume River and seized 90 percent of the Citadel including the Imperial Palace. The red and blue banner with the gold star of the *Viet Cong* was soon seen flying in the Citadel flag tower. Once the city was secured, the communists established their own civil government and rounded up known government officials, sympathizers, and foreigners including American civilians and military personnel. It was clear they intended to stay.[6] (See Map 10.)

Just Holding On

Truong's foresight had saved his command from total disaster. By 4 a.m. the Black Panther (Hoc Bao) Company had successfully blocked the *6th NVA* at the Tay Loc Airfield. The Black Panthers, reinforced by the 200-man division staff, were later successful in recapturing the medical company's cantonment area. Most importantly, the 1st ARVN Infantry Division's command and communications structure remained intact. Meanwhile, the attack by the *4th NVA* in the south had isolated the MACV compound, but failed to take it in spite of repeated assaults. Except for these two small pockets and the retention of the Landing Craft Utility (LCU) ramp, the situation was bleak. However, the failure of the *NVA* to capture these two strongholds and close the river traffic permitted the South Vietnamese and Americans to bring in reinforcements and eventually mount a counteroffensive.[7]

Although the situation was not clear to the senior American and ARVN commanders, they dispatched forces within hours to relieve the embattled defenders of Hue. The 1st US Marine Division committed units of the 1st Marine Regiment, including a platoon of M48 tanks from the 3d Battalion,

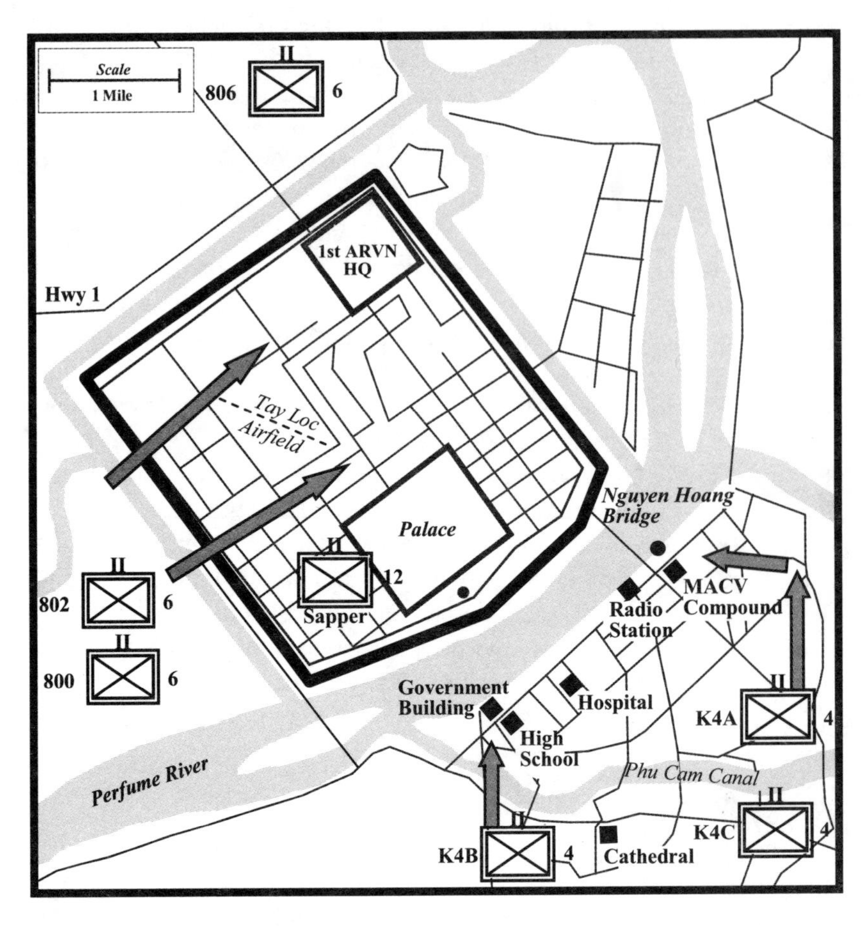

Map 10. North Vietnamese attack, 31 January 1968.

to this effort. Truong ordered his 3d Regiment, 1st Airborne Task Force and elements of the 7th Armored Cavalry, equipped with light tanks and M113 armored personnel carriers, to the Citadel. The heavy armor of the Marine M48 tanks had been en route from Phu Bai to the LCU ramp at Hue for embarkation and transfer north to the 3d Marine Division at Dong Ha. Arriving rather piecemeal, these reaction forces met a stubborn resistance by the *NVA* outside the city, but managed to break through to the MACV compound. An attempt by the US Marines to cross into the Citadel on 31 January failed against the well-fortified *NVA* positions. The situation remained critical for three days as three more Marine companies, two Marine battalion command groups, and a Marine regimental command group under the command of Brigadier General Foster "Frosty" C. LaHue arrived in the compound.[8]

The tale of the relief columns to reach the embattled defenders of Hue was a prelude to the tough fighting to come. The initial relief consisted of two battalions of the 3d ARVN Regiment moving east along the northern bank of the Perfume River, while the airborne battalions and a cavalry troop sought to fight their way into the 1st ARVN Division headquarters compound in the northeast corner of the Citadel. The 3d ARVN Regiment was forced to fall back under heavy fire, and two battalions were surrounded and held under heavy fire. The 1st ARVN Infantry Battalion was able to extract itself from its predicament and reach the Citadel the next day via motorized junks on the river. After fighting for several days, the 4th ARVN Battalion was finally able to break free. Meanwhile, a squadron from the 7th ARVN Armored Cavalry Regiment attempted to break through, but came under heavy fire, including B-40 rocket grenades, forcing it to halt. Reinforced by US Marines and M48 tanks, the 7th ARVN armored column crossed the An Cuu Bridge into the new city. Pushing toward the central police headquarters in southern Hue, the ARVN cavalrymen attempted to relieve the besieged policemen. When a direct hit from a *NVA* B-40 rocket killed the squadron commander in his tank, the 7th ARVN armor fell back.[9]

As the Marines of Company A, 1st Battalion approached the southern suburbs of the city, they came under sniper fire and B-40 rockets. The Marines dismounted from their trucks and cleared the houses on both sides of the main street before proceeding. The convoy crossed the An Cuu Bridge, but was immediately caught in a murderous crossfire, forcing them to dismount again. At this moment a B-40 rocket struck and killed the commander of the lead tank, causing the advance to falter. The only reinforcements available were the command group of the 1st Battalion and Company G from the 5th Marines. Under the command of Lieutenant Colonel Marcus J. Gravel, this ad hoc force reached Company A in the afternoon. The M48 tanks were kept on hand to provide direct fire support to the Marines, but the trucks were used to evacuate the dead and wounded. Once the Marines were consolidated and the few surviving 7th ARVN M41 Walker Bulldog tanks rounded up, the advance continued with the armor in the lead. To the great relief of the defenders, this force arrived at the MACV compound around 1500.[10]

While the rest of the Marine task force proceeded toward the Citadel, Company A stayed behind to regroup and guard the MACV compound. The M48 tanks were too heavy to cross the river bridge into the old city and were instead positioned to provide direct fire support. The accompanying ARVN M24 Chaffee tank crews refused to lead the assault across the bridge, leaving that dangerous task to the Marine riflemen. As the Marines started across in small groups, the *NVA* sprayed the bridge and its approaches with heavy automatic and recoilless rifle fire. Two platoons were successful in crossing, but were forced to withdraw. Nearly a third of the attacking Marines were dead or wounded in the effort. By 2000 the fighting halted as both sides consolidated their positions and prepared for the next day.[11]

Senior ARVN and American leaders conferred in Da Nang about the situation around Hue. Major General Hoang Xuan Lam, the I Corps Commander, and Lieutenant General Robert Cushman, Commanding General, III Marine Amphibious Force (MAF) set unit objectives in good faith but without knowledge of the actual situation. In deference to national pride, it was decided the South Vietnamese would be responsible for the liberation of the Citadel while the Americans would continue to clear the city south of the river and cut enemy communications to the west. In the hope of sparing the old city from massive damage, it was also decided not to employ heavy artillery or fixed-wing aircraft against targets in the Citadel. Hue was a far more difficult objective than these senior commanders expected.[12]

On 1 February, the South Vietnamese forces initiated their operations to clear the enemy from inside the Citadel. There was some success. The 2d and 7th ARVN Airborne Battalions, supported by armored personnel carriers and the Black Panther Company, were successful in recapturing the Tay Loc Airfield inside the walls. By 1500 the 1st Battalion, 3d ARVN reached the 1st ARVN Infantry Division command post. However, the 2d and 3d Battalions of the 3d ARVN Regiment were unable to penetrate the enemy defenses.[13]

Meanwhile, at 0700 on 2 February the composite 1st Marine Battalion launched a two-company assault supported by tanks. The direction of the attack was toward the jail and the Thua Thien provincial headquarters building and prison, a distance of six blocks west, with the additional objective of fully securing the LCU ramp. The *4th NVA Regiment* was positioned in force and waiting. The lead elements did not get a block away from the MACV compound before they came under heavy fire. One of the M48 tanks received a direct hit from a 57mm recoilless rifle that disabled the vehicle and wounded the crew. This stopped the attack, and the Marines withdrew to the MACV compound to regroup. The damaged tank was quickly repaired and the crew replaced by riflemen.[14] (See Map 11.)

After receiving an update on the situation, LaHue realized the enemy strength in Hue was much greater than originally estimated. To streamline

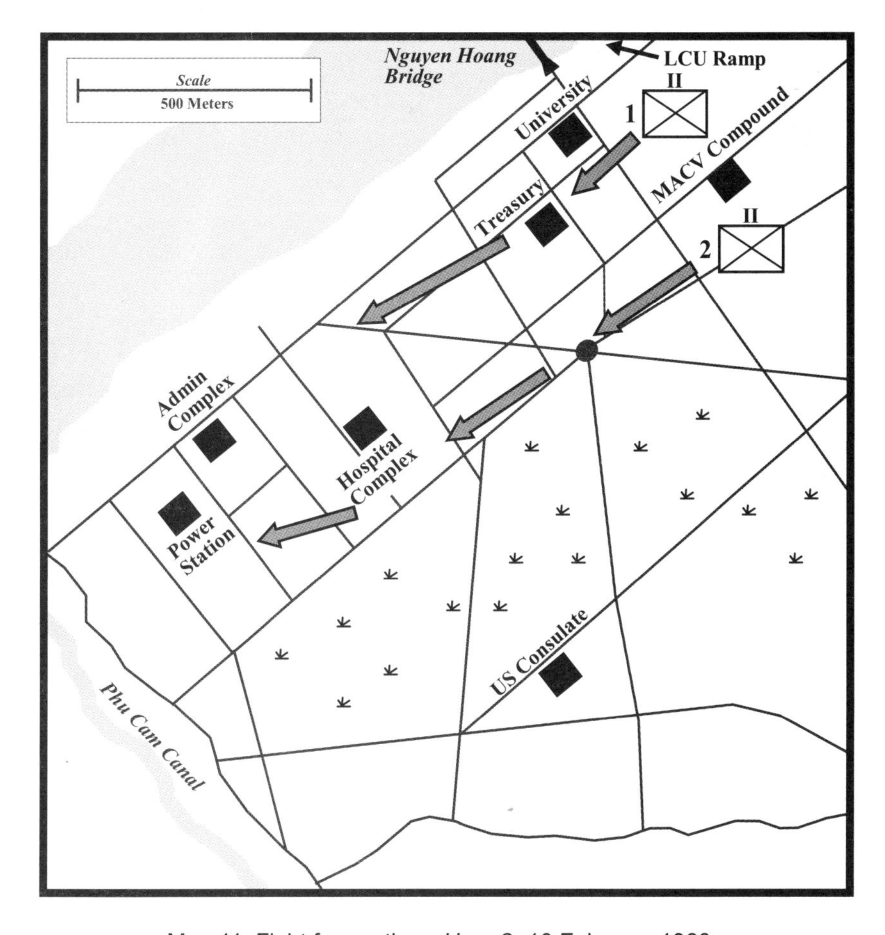

Map 11. Fight for southern Hue, 2–10 February 1968.

the chain of command, he gave Colonel Stanley S. Hughes of the 1st Marines full tactical control of the forces in the southern city. Hughes then promised Gravel of the 1st Battalion additional reinforcements. By late afternoon, Company F, 2d Battalion, 5th Marines arrived via helicopter at the MACV compound. Marine CH-46 helicopters also airlifted about half of the 4th Battalion, 4th ARVN Regiment onto the recently secured airfield in the Citadel. All of these reinforcements came in during poor weather conditions that deteriorated enough to cancel flights before the airlift was complete. Also, *NVA* forces attempted to disrupt the operation with mortar and sniper fire opposed the landings. Tasked to relieve an isolated US Army signal relay station in the Marine sector, the newly arrived Company F could not break through to it. The fighting again ended at nightfall with little to show for it. Ordered to hold their positions, the Marines prepared themselves for the next day.[15]

Pushing Back

The weather conditions that inhibited airlift operations on 1 February declined the following day. A constant drizzle developed that occasionally increased to a cold drenching rain. Temperatures dropped into the 50s, which was quite cool for Vietnam. The low ceiling and poor visibility prohibited air support and airlift operations and inhibited fire support missions. Although artillery and naval fire support was in range, the forward observers had difficulty sighting targets and adjusting fires in the miserable weather. In spite of the weather, the Marines were able to make some gains on 2 February. They relieved the signal relay station and reached the university campus after a savage firefight. Unable to get further reinforcements to Hue by air, the Marines resorted to sending them in by ground. To get there, they had to run a gauntlet of *NVA* fire.

The US Army was also on the move toward Hue as the 3d Brigade, 1st Cavalry Division advanced from the west to establish blocking positions and to cut off the *NVA* forces in the city from outside support. Within two days, the 2d Battalion, 12th Cavalry Regiment established positions on the high ground with excellent observation of the main enemy routes in and out of Hue. From there it was in position to interdict all daylight movement of *NVA* forces by using artillery fires. The 5th Battalion, 7th Cavalry Regiment conducted search operations along enemy routes further west. The success of these forces to restrict *NVA* forces from moving through the area was limited due to the poor weather conditions and low ceiling. The North Vietnamese continued to pour reinforcements and supplies into Hue.[16]

Although *NVA* sappers had blown the railroad bridge west of Hue, they had left the bridge over the Phu Cam Canal intact. It was a costly mistake.[17] Company H of the 2d Marine Battalion used this key bridge to reach the MACV compound. Two US Army M55 trucks mounting four quad .50-caliber machine guns, commonly called Dusters by the Marines, escorted the convoy. Additionally, there were two M50 Ontos light armored vehicles each mounting six 106mm recoilless rifles.[18] Encountering stiff *NVA* resistance as they approached the city, the convoy escort applied heavy firepower and made it safely to the compound. By the end of the day, the Marines had sustained 2 dead and 34 wounded and claimed to have killed 140 enemy.[19]

The commander and staff of both the 1st and 2d Battalions from the 5th Marines arrived on 3 February. Another attack against nearby *NVA* positions was ordered, this one composed of two companies supported by tanks with one company held in reserve. Company frontages were about a block in width, but there were not enough riflemen to do the job.

The buildings in this area proved impervious to small arms fire and to the M72 Light Antitank Weapons (LAW) carried by the Marines. This assault soon stalled, but the forward positions gained were held. To the east of this action, Company A of the 1st Marines captured an abandoned South Vietnamese police station and a number of small arms. It avoided entering the International Control Commission (ICC) building under the terms of the 1954 Geneva Accords, containing and bypassing it instead.[20]

Fighting resumed the next morning with the Marines continuing their efforts to push forward and clear the city of *NVA* soldiers. The fighting was building-to-building, room-to- room against a determined enemy. The *NVA* fought from every crevice possible, and secreted themselves in the eaves and ceilings. The *NVA* soldiers put up a stiff resistance at the church, prompting a very reluctant Marine commander to authorize destruction of the structure. As the battle for Hue raged and the casualties mounted, the qualms against collateral damage almost disappeared.[21]

A twist in the nature of fighting became apparent as the battle in Hue developed. In the jungles of Vietnam, the *Viet Cong* and *NVA* typically employed hit-and-run tactics and ambushes and then melted away. Often, Marines and soldiers served their entire tours fighting an elusive enemy without clearly seeing him. At Hue it was different. There the *NVA* dug in and showed no signs of leaving. Many Marines reacted with enthusiasm for the opportunity to kill as many of their foe as possible.[22]

Trained to fight in the jungles of Vietnam, the young Americans began to adapt to the tempest of urban terrain. Marines initially used smoke grenades to cover their movements, but the *NVA* simply sprayed the streets with automatic weapons fire. Learning quickly, the Marines shifted tactics and used the smoke grenades to provoke their enemy into firing. Then tanks or Ontos vehicles engaged these targets. The riflemen then crossed the open zones using the armored vehicles, smoke, and the dust that was raised by the back-blast as cover. Additionally, Marines developed the method of using explosives or a 106mm recoilless rifle to breach a wall thus allowing a fire team to rush through the gap. In one instance, the Marines assembled a recoilless rifle inside a building to get at a particularly troublesome *NVA* position. To avoid the deadly overpressure, the Marines fired the weapon using a long lanyard that led outside. It was fortunate they did so because the building collapsed when the weapon discharged. It destroyed the target, but the gun was still operational once it was dug out of the rubble. Mortars were also used extensively to blow open buildings for the riflemen and to destroy targets in areas inaccessible to direct fire.[23]

These tactics proved effective; however, they did not always work. In the fight for the Treasury Building, the thick walls and steel door remained impervious to tank, recoilless rifle, and mortar fire. Again, the Americans turned to improvisation. To break the impasse the Marines scrounged a number of E8 rocket launchers designed to fire CS tear gas from the MACV compound. These weapons could lob 35mm gas grenades up to 250 meters. Within minutes, a thick cloud of tear gas filled the Treasury Building. Donning their M17 protective masks, the Marines of Company F assaulted the facility under the covering fire of mortars and machine guns. The remaining *NVA* forces beat a hasty retreat.[24]

The M48 tanks were a welcome addition to the Marine arsenal, but they had severe limitations in Hue. The streets were very narrow and a B-40 rocket could, and often did, appear from nowhere. The tanks drew fire from all *NVA* weapons, including small arms, mortars, and rockets. One of the M48 tanks reportedly was hit by rockets more than 120 times and went through at least six crews. In each case, the tanks were recovered, repaired, and put back on the line within hours. Incidentally, the Marines were much more enthusiastic about the M50 Ontos vehicle than the medium tank. The 106mm recoilless rifle was an effective weapon, very familiar to the rifle-men. The thin armor was sufficient only against small arms, but the small size of the M50 allowed it to maneuver down any street.[25]

The Marines also had to modify their command and control procedures as they adapted to fighting in the city. The standard 1:50,000 scale maps proved inadequate in scale and detail for the operation. The local government-made maps depicted all of the major buildings and assigned them a specific number in the map key. The Marines simply set up a control system corresponding to the building number system on the map, and procured and issued a small number of maps to unit leaders. This system had a serious drawback in that artillery spotting still required use of the 1:50,000 scale tactical maps. Thus, a great deal of care was needed to avoid fratricide. As the poor weather precluded most artillery target acquisition missions, this liability was not fully manifested in the battle. The hasty issue of the 1:2,500 scale maps somewhat relieved this map situation, but there were only enough for the battalion staffs and one per company.[26]

While the Marines in the new city adjusting to urban warfare made slow progress, the ARVN attack in the Citadel ground to a halt. Although the northeast corner of the city and the airfield were secured, the 1st and 4th ARVN Battalions were halted about halfway to their day's objective. Another setback occurred as North Vietnamese sappers destroyed the Nguyen Hoang Bridge over the Perfume River and the An Cuu Bridge over the Phu Cam Canal. This closed the only land route to the southern half of the city. Now supplies and logistics could only be brought in by air or via the LCU ramp. With the deteriorating weather, the LCU ramp became an object of major importance and the narrow ship channel and proximity of *NVA* forces made it a very precarious line of communications. Fortunately for the Marines, a large amount of supplies were brought across the bridge prior to its demolition.[27]

Both Marine battalions continued to see heavy fighting over the course of the next several days. Continuing to use M48 tanks and M50 Ontos vehicles in support, the Marines were successful in capturing the hospital and the prison. The Marines were learning the tactics and value of entering a building from the top and working downward, and used this technique whenever possible. The high use of CS gas forced the use of M17 protective masks, which greatly restricted peripheral vision and depth perception. By using all weapons available and improvised tactics, the Americans were able to secure their objectives.[28]

In the Citadel, Truong adjusted the disposition of his troops. He shifted the 4th ARVN and temporarily relieved the three airborne battalions. The 4th Battalion was sent to secure the airfield and push forward to the southwest wall. The 1st ARVN Battalion found success in recapturing the An Hao Gate, located in the northwest corner of the old city. The remaining three battalions of the 3d ARVN Regiment were committed to clearing the Citadel to the southeast wall. Fighting was tough, but the ARVN forces were at least making slow progress.[29]

An important milestone in the battle for Hue occurred on 6 February as the Marines captured the provincial headquarters in the new city. Expending over 100 mortar rounds and a large amount of other heavy ordnance, Company H overwhelmed the dogged defenders. Two M48 tanks supported this attack, one of which took multiple hits from B-40 rockets but continued to fire. The headquarters building had become a symbol for both sides, and the Marines lost no opportunity to magnify that image. Immediately after the capture, two Marines sprinted to the flagpole and ran up the American flag. As there was no South Vietnamese flag present, this act was in direct violation of protocol, but few if any Marines cared. As it turned out, the provincial headquarters was more than a symbolic victory. It had been used as the command post for the *4th NVA Regiment* and with this loss the intensity of the fighting diminished in the modern city.[30]

The fight was far from over though. Using the cover of darkness and poor weather, the North Vietnamese were able to evade the screen of the 1st Cavalry Division and bring a large number of fresh troops into the

Citadel. On the night of 6 February and into the next morning, the *NVA* launched a savage counterattack against the 2d Battalion, 4th ARVN, forcing it to fall back to the airfield with heavy losses. The *NVA* was checked when a fortunate break in the weather allowed the South Vietnamese air force to strike targets in the old city. Truong then redeployed the three battalions of the 3d ARVN Regiment to the headquarters compound, moving them to the north side of the Citadel by employing motorized junks. By the end of 7 February, the ARVN forces in the Citadel consisted of two armored cavalry squadrons, the 3d ARVN Regiment, four airborne battalions, a battalion from the 4th ARVN Regiment, the Black Panther Company, and a company from the 1st ARVN Regiment. However, these forces were armed with rather antiquated infantry weapons and suffering from heavy casualties. They were up against a well-fortified, tough opponent who still controlled over half of the Citadel. The ARVN forces would make little headway over the next several days. The only good news was that the Black Panther Company was successful in retaking the airfield.[31]

On the following day, Truong revised his battle plans. The arrival of the lead elements of a South Vietnamese marine task force allowed him to plan for the relief of the battered airborne units, which were then to be redeployed to the fighting around Saigon. Unfortunately, weather conditions restricting helicopter operations delayed these South Vietnamese marines for three days. Desperate for men and firepower, Truong then requested a US Marine battalion to participate in the battle for the Citadel.[32]

Battle of the Citadel

The US Marines were successful in securing the southern half of Hue by 10 February. With that half of the city secured, the Marines could then focus on assisting the embattled ARVN forces in the Citadel. This task was difficult, as most of the bridges into the walled city had been destroyed. The newly arrived 1st Battalion, 5th Marine Regiment was designated to operate within the northeast corner of the Citadel. By midday on 11 February, the Marines arrived in force at the Citadel airfield using CH-46 helicopters. Company A, with five M48 tanks attached from the 1st Marine Tank Battalion in support, arrived in a LCU and entered the old city from the south. These Marines relieved the ARVN airborne task force, which soon departed the city. Two 4.2-inch heavy mortars and a 105mm howitzer battery were positioned close to the city to provide indirect fire support for the coming battle.[33]

The buildup of ARVN and American forces was not lost on the *NVA*, which chose this time to redouble its efforts to hold Hue by sending in reinforcements and launching savage attacks. Characterized by the brutal and bloody street actions of urban warfare, the battle for the Citadel raged from 11 to 25 February. Again fighting was block-by-block, building-to-building, and room-to-room. Completely gone were the efforts to limit collateral damage as heavy artillery and naval gunfire were used to support the ground actions.

The US Marines began the offensive to clear their zone on the morning of 13 February, using the same techniques as the previous weeks. Two companies were to advance abreast clearing each building encountered, while one company was held in reserve. Tanks and Ontos vehicles were used as direct fire support. Unaware of the withdrawal of the ARVN paratroopers, the Marines found themselves up against a host of *NVA* soldiers. With tanks in the lead, the Americans moved forward but did not even reach their line of departure. The *NVA* had constructed spider holes along the old walls and occupied other positions covering the approaches. A hail of automatic fire, grenades, B-40 rockets, and mortars stopped the Marines. The best news of the day was the erection of a pontoon bridge across the Perfume River to replace the damaged An Cuu span, allowing the shipment of food and supplies into the old city.[34] (See Map 12.)

The next day, 14 February, the Marines employed their artillery and naval fire support as a walking barrage ahead of their advance to soften up the *NVA* defenders. The generally flat trajectory of the guns limited their effectiveness in the urban environment though. An unexpected break in the weather during the afternoon allowed F-4 Phantoms and F-8 Crusader fighter-bombers to fly support missions, but their overall effect turned out to be minimal. The Marine attack stalled again under the withering fire from the *NVA* positions. The ARVN forces fighting in the southeast corner of the old city fared no better. In fact, an *NVA* counterattack managed to cut off the 1st Battalion of the 3d ARVN Regiment, requiring two days of hard fighting to relieve these men.[35]

On 15 February, the Marines showed some progress by capturing a tower along the wall in their sector. Aided by A-4 fighter-bombers during another break in the weather, the Americans took and held the structure after hand-to-hand fighting and heavy casualties. The battalion pushed forward the next day, gaining a few more buildings. It was generally felt that the *NVA* in this part of the city fought better and employed more sophisticated tactics than those encountered south of the river. The *NVA* in the Citadel had dug trenches and spider holes, used roadblocks, and did not hesitate to counterattack to regain key positions. With the end of fighting on 16 February, the Marines in the northeast corner of the Citadel paused to regroup, rearm, and resupply. All men were in need of rest and the armored vehicles required refueling and ammunition.[36]

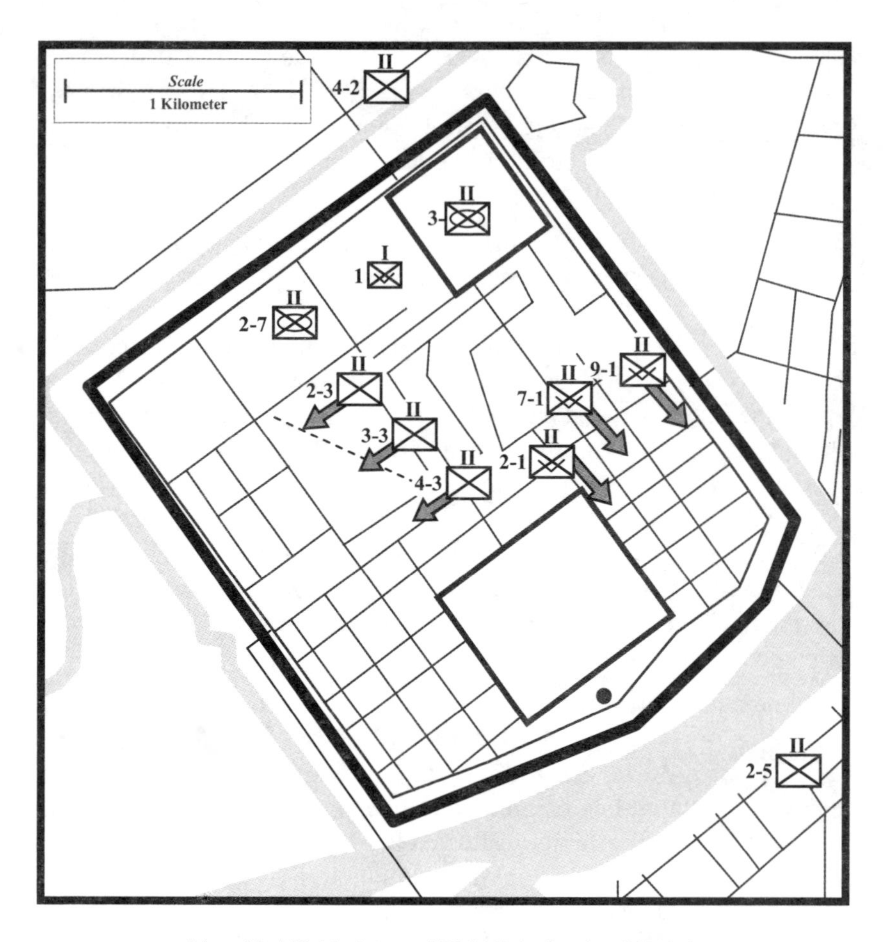

Map 12. US Marines—ARVN fight for the Citadel.

With the poor weather and limited effect of heavy artillery against the stone structures, the Marines had to rely on their organic weapons, particularly the mortars and armor. The tanks and Ontos vehicles were attached down to platoon level to increase their responsiveness. While the riflemen provided cover, the armor engaged point targets as needed. Often, the vehicle commanders would dismount and go forward with the riflemen to reconnoiter. On engaging a target, the armored vehicles would quickly reverse to gain cover and the Marines would surge forward. The tank crewmen quickly discovered that the standard high-explosive rounds did little damage to the stone or masonry walls in the old city. The rounds often ricocheted back into friendly troops. The tank crewmen switched to high explosive antitank (HEAT) rounds, which usually could breach the ancient walls with four or five rounds. Casualties were high among the armor crewmen, but the M48 tanks generally stood up well against the

B-40 rockets. The tanks were soon back in action with replacement crewmen. Another key weapon in the fight was the 4.2-inch mortar firing CS shells. The *NVA* was generally ill equipped to deal with tear gas and often abandoned their positions once the concentration became high.[37]

A noted success during the night of 16 February was the interception of a North Vietnamese transmission ordering a battalion to cross the canal into Hue from the west. Accordingly, the Marines and ARVN indirect fire systems massed their fire on the bridge. A later intercept revealed that a high-ranking *NVA* officer, possibly a general, and many men had been killed or wounded. The request by his successor to withdraw was denied; the new commander was directed to continue the fight. This incident and other indications confirmed that the *NVA* were reinforcing Hue by night. To tighten the screen, the 1st Brigade of the 101st Airborne Division was deployed to the west of Hue to augment the efforts of the 1st Cavalry Division. Over the next week, these forces pushed closer to Hue and became more successful in cutting the *NVA* lines of communication.[38]

At his headquarters, Truong made plans for the final battle for the Citadel. The Vietnamese marine task force, now consisting of three battalions, was to push forward and clear the southwestern wall. The 3d ARVN Regiment was tasked to drive through the center of the city south toward the Imperial Palace. The 1st Battalion of the 5th US Marines would continue its operations in the southeast sector of the old city, near where they had landed on the LCU ramp a few days earlier. The Marines in this sector had been relatively inactive since their arrival, having experienced shortages in food and ammunition. Rounds for the tanks and Ontos were in critically short supply due to the loss of an LCU by enemy fire.[39]

The pause in offensive operations ended the morning of 21 February as the 1st Battalion of the 5th Marines surged forth again. This time there was a variation in the tactics used. In a rare maneuver, three ten-man teams launched a night attack on a key two-story building and two other buildings that covered the flank of the *NVA* positions. The surprise was total, and the teams occupied the buildings almost unopposed. As the unaware *NVA* soldiers moved to reoccupy the buildings in the early morning, a hail of automatic fire from the Americans met them. Meanwhile, the rest of the 1st Battalion rushed forward to occupy the buildings. The stunned *NVA* fell back to their subsequent positions and defended with their usual tenacity. Unknown to the Americans, during an *NVA* counterattack several high-ranking officers and political leaders used the diversion to slip out of Hue. This marked the beginning of the end.[40]

At dawn on 22 February, the Marines of the 3d Battalion in the southeast corner of the city pushed forward. To their surprise, they encountered only an occasional sniper or mortar round, as their enemy had seemingly disappeared. The riflemen pushed to the southern wall where they hoisted an American flag and proceeded to the southern gate. Armored vehicles provided direct fire support as before, engaging known and suspected *NVA* positions. The ARVN and South Vietnamese did not have an easy time of it in their sectors and still encountered fierce resistance. That evening all American and South Vietnamese forces shifted to the defense and prepared for another fight in the morning.[41] (See Map 13.)

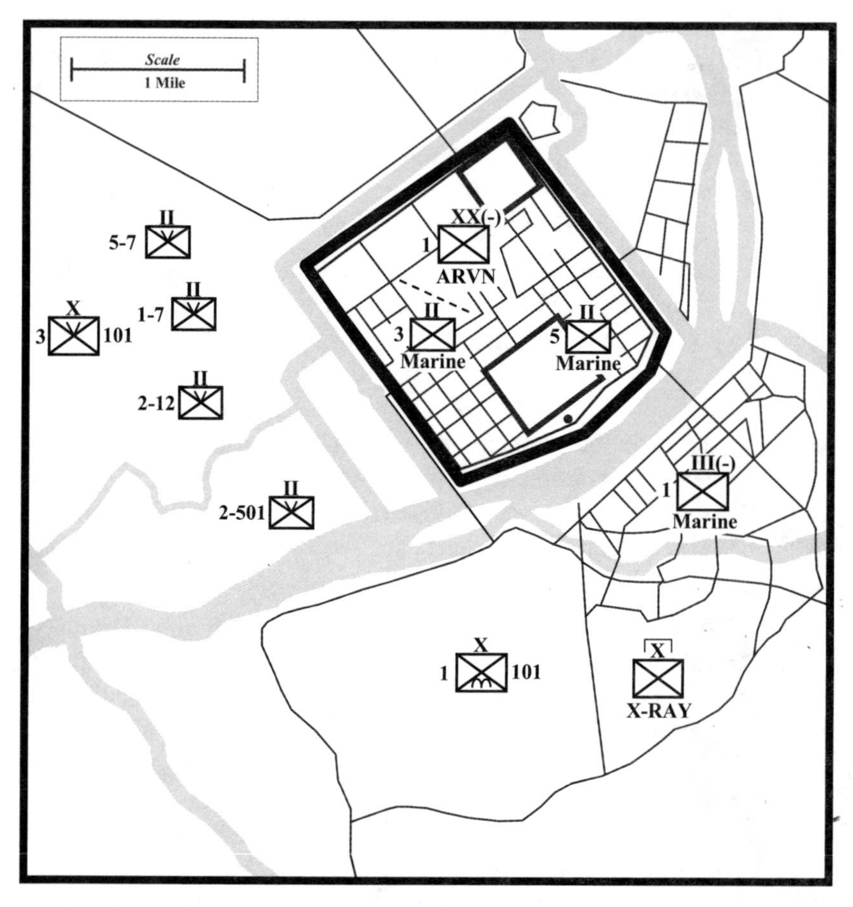

Map 13. Tactical dispositions, 24 February 1968.

The next morning, the 1st Battalion of the 5th Marines was able to secure its sector in the southeast corner of the old city in what would be its last major action. The Marines had hoped to participate in the liberation of the Imperial Palace, but this task was reserved for the South Vietnamese as a matter of national pride; however, US Marine armor was positioned to provide direct fire support. The bulk of the 1st Battalion was relieved in the line by ARVN forces and moved to the north of the city to join the 2d Battalion in securing that sector. That evening, determined NVA counterattacks were repulsed and the 3d ARVN Regiment mounted a surprise night attack that knocked their enemy off balance. The ARVN forces did not let up and were successful in capturing the Citadel flag tower in the morning of 24 February. There, the flag of South Vietnam finally replaced the *Viet Cong* banner. The Black Panther Company continued its attack and secured the Imperial Palace by the late afternoon. By early morning of 25 February, the 4th Vietnamese Marine Battalion eliminated the last *NVA* strongpoint in the southwest corner of the old city. Except for a week of mopping up operations outside the city, the battle for Hue was over.[42]

In Retrospect

The recapture of Hue was a particularly bitter fight involving house-to-house fighting and heavy casualties, extensively damaging about 80 percent of the city, and leaving over 110,000 people homeless. Approximately 5,000 civilians were killed, including about 3,000 who were executed by the *Viet Cong* and *NVA* and buried in mass graves. US assistance agencies helped the South Vietnamese authorities in restoring order to Hue and stopping the widespread looting.[43]

The losses for the combatants were high as well. The *NVA* and *Viet Cong* forces lost at least 2,000 to 5,000 dead. The South Vietnamese lost over 300 killed and about 2,000 wounded. In the fighting to cut off the *NVA* forces to the west of the city, the 1st Cavalry Division (Airmobile) lost 68 killed and 453 wounded, while the 101st Airborne Division reported 6 dead and 56 wounded. The three US Marine battalions fighting in Hue sustained total casualties of 142 dead and nearly 1,100 wounded, the losses being very high for unit leaders. Lieutenants were left in command of companies and corporals in charge of squads. Combined, the allied unit casualties totaled over 600 dead and 3,600 wounded.[44]

Armored forces were a key element in the hard-won victory at Hue. Tanks brought the traditional firepower and mobility to the battle with heavy enough armor to protect their crewmen from most of the *NVA*'s weapons. The M48 tanks in particular were able to absorb a huge amount of punishment and keep fighting. This ability to withstand damage allowed the M48 tanks to fight throughout the battle after makeshift repairs and rotating crews. Fortunately the fire control, loading, and driving were simple enough to allow inexperienced personnel to ride tanks into combat when no replacement armor crewmen were available.

The lighter M50 Ontos vehicles lacked suitable armor and were vulnerable to enemy fire. A hit from a B-40 rocket grenade would easily render the crew as casualties and disable the vehicle. The gunner and loader had no armor protection except the bulk of the vehicle in front of them. They were able to bring their weapons to bear only when used in very close cooperation with the accompanying riflemen. However, because the typical Marine was familiar with the 106mm recoilless rifle and the small Ontos could maneuver in the narrow streets, these vehicles were in high demand.

The importance of the armor protection of combat vehicles was clear in the narrow streets of Hue, particularly the Citadel, where the ability of the tanks and Ontos vehicles to maneuver and target effectively was restricted. The row-on-row of one-story, thick walled houses jammed together with narrow streets was a tanker's nightmare, with unlimited opportunities for a B-40 gunner to fire from any angle. The Marine riflemen could not protect all the vehicles all the time in such an environment, and a hit on one was inevitable. It was vital that these vehicles could take a hit and keep fighting.

Procedurally, the use of armor in Hue was ad hoc in nature as there was no practiced urban doctrine in either the Marine or ARVN forces. Crews and accompanying riflemen learned as they went along. Armor was used primarily in the support role for the infantry, but on occasion it led limited advances with the Marines crouched behind. Resupply and maintenance was done by withdrawing the vehicles to the rear, and not by any set schedule or plan. Combat was done by day as night fighting by the Americans was rare in this era; the M48 tanks had a rudimentary night sight while the Ontos vehicles had none. For the riflemen, the few available night vision devices were heavy and cumbersome and the batteries had a short operating life. Artillery, mortars, flareships, and gunships usually provided illumination. This illumination was very much weather-dependent, and the weather during the battle for Hue was generally poor.

Although a long and bitter fight, the battle for Hue demonstrated the ability of armored forces to move under heavy fire and to bring even heavier firepower to their enemy, even in urban terrain.

Notes

1. Donn A. Starry, *Armored Combat in Vietnam* (New York: Arno Press, 1980), 54–55, 66–77, 115. Bryan Perret, *Iron Fist: Classic Armoured Warfare Case Studies* (London: Arms and Armour Press, 1995), 189–190, 191. R.P. Hunnicutt, *Patton: A History of the American Main Battle Tank* (Novato, CA: Presidio Press, 1984), 373, 381. The South Vietnamese adopted the M113 armored personnel carrier and M24 and M41 light tanks for their armored cavalry regiments.

2. Perret, 189–190. Hunnicutt, 225. The B-40 rocket propelled grenade, a variant of the Soviet-made RPG-2, was notorious for its inaccuracy, due to either the weapon or the operator. On average, only one in seven hit its intended target. It could penetrate just over seven inches of armor, but it needed to hit a flat surface at a nearly 90-degree angle to detonate. The sloped sides of the M48 tank made this very difficult. The range of the weapon was rated to 150 meters, but in the streets of Hue it was much less.

3. Keith W. Nolan, *Battle for Hue: Tet, 1968* (Novato, CA: Presidio Press, 1983), xii–xiii, 3–4. The Perfume River is also known as the Huong River.

4. Eric M. Hammel, *Fire in the Streets: The Battle for Hue, Tet 1968* (Chicago, IL: Contemporary Books, 1990), 7–8, 11, 16, 22. Nolan, 3–4, 121. Brigadier General Truong was the only senior field commander, ARVN or American, to prepare any sort of defense in the Hue area before the onset of the Tet Offensive. He was considered by many to be one of the best senior combat commanders in the ARVN. William C. Westmoreland, *A Soldier Reports* (New York: Da Capo Press, 1976), 310–311. The Tet holiday is the most important holiday for the Vietnamese. A rough comparison of the Tet holiday is rolling Christmas, Independence Day, and Labor Day into one.

5. Don Oberdorfer, *Tet!* (Garden City, NY: Doubleday & Company, Inc., 1971), 121. This source primarily chronicles the suffering and death of the thousands of South Vietnamese during the battle. Mr. Oberdorfer was a journalist for the *Washington Post* during the battle.

6. Nolan, 9–10. Hammel, 29–31, 37–40, 43, 131.

7. Many LCUs were former LCTs (Landing Craft, Tanks) from World War II that were refurbished and employed in Vietnam as a logistical workhorse for the US Navy. Most were left behind during the American withdrawal and presumed to have been scrapped, but there may be one or two still around.

8. Hammel, 93–94. Brigadier General LaHue had just established his command post at Phu Bai on 13 January and had little time to become acquainted with his new area of operations.

9. Hammel, 93. Perret, 189. The ARVN forces were equipped with the M24 Chaffee and M41 Walker Bulldog light tanks and M113 armored personnel carriers. The standard infantry weapon at the time for the ARVN was the M1 Garand rifle from World War II.

10. Hammel, 63, 96–98, 160. Nolan, 18–19.

11. Hammel, 83. Nolan, 19–20.

12. Nolan, 21, 46.

13. Ibid., 27–29.

14. Hammel, 104–106.

15. Hammel, 59, 113. Nolan, 31. The poor weather would last for three weeks.

16. Hammel, 190–193, 308–310. Nolan, 28. Westmoreland, 329. General William C. Westmoreland, Commander of US Forces in Vietnam, was formulating a broad plan to reinforce I Corps and prepare for mobile operations aimed at cutting off the *NVA* forces in Hue, Saigon, and Khe Sanh. Most of the fighting during the Tet Offensive ended by 11 February, but Hue lasted almost two weeks longer.

17. Hammel, 48, 77. Nolan, 29. Senior *NVA* field commanders candidly conceded that the seeds for eventual defeat were sown by the *4th NVA Regiment's* inability to secure the southern half of the city and prevent reinforcement to the MACV compound and the embattled 1st ARVN Division headquarters in the Citadel.

18. Hammel, 141. The Ontos had many drawbacks. The recoilless rifles were mounted externally, requiring the crew to operate them in the open. The armor was not sufficient to defeat bullets of most calibers, and the engine was powered by volatile gasoline.

19. Hammel, 85–87, 116. These armed convoys were often called roughrider convoys. Not all of the armed trucks in the convoy were actually Dusters, which referred to the Army M42 tracked vehicle mounting dual 40mm antiaircraft guns. The Marines called both the M55s and M42s by that name. For a brief but comprehensive overview of convoy operations in Vietnam, see Richard E. Killblane, *Circle the Wagons: The History of US Army Convoy Security* (Fort Leavenworth, KS: Combat Studies Institute Press, 2005).

20. Hammel, 147–149. Nolan, 44.

21. Hammel, 139.

22. Nolan, 56–57, 88. An example of the high morale and desire to eliminate the enemy was conveyed in almost all of the wounded Marines, who expressed a desire to get back to their units.

23. Hammel, 264, 301. Nolan, 38, 141.

24. Hammel, 135–136. Nolan, 51.

25. Starry, 116. Hammel, 152, 301. Nolan, 108, 141, states that a flame-throwing tank variant was in Hue, but that its tanks were empty. No other account confirms this.

26. Each battalion received three maps, one for the commander and staff and one per company. The Army Map Service produced these.

27. Hammel, 187, 239. Nolan, 38. The Marines called the Nguyen Hoang Bridge the "silver bridge" because it was painted metallic silver.

28. Nolan, 43.

29. Hammel, 190.

30. Hammel, 236–237, 251. Nolan, 74–77, 81. A CBS news crew filmed the event, comparing it to the flag raising on Iwo Jima in World War II.

31. Hammel, 254–255, 261, 302. Allied intelligence services initially

identified the *4th, 5th,* and *6th NVA Regiments* operating in and around Hue. Later these organizations confirmed the presence of elements of the *29th, 90th*, and the *803d NVA Regiments*. Intelligence estimates surmised approximately 18 NVA battalions fought in the city and surrounding areas, not including *Viet Cong* units. This represented 8,000 to 11,000 troops.

32. Hammel, 190–191, 261–263. Nolan, 118–119, 122–123. The ARVN cavalry units also took heavy casualties. Eight of 12 M113 APCs were lost in short order and the morale of the troops was low. The M113s were presumably used as mobile carriers for their .50-caliber machine guns, but their thin armor was no match for the B-40 rockets.

33. Hammel, 220–221, 262, 275. The initial positions of the Marine artillery to the south made firing into the Citadel a hazardous affair. If the rounds went long, as they often did in manual spotting, they would land on friendly positions. Shifting the artillery shortened the range and eliminated this hazard.

34. Hammel, 270.

35. Hammel, 271–273. Nolan, 140–141.

36. Hammel, 281.

37. Nolan, 51, 90. Some *NVA* officers and noncommissioned officers had protective masks, but the rank and file did not. The use of CS gas was very demoralizing to the *NVA*.

38. Hammel, 295. Nolan, 137.

39. Nolan, 122–123, 146, 149, 157.

40. Wilbur H. Morrison, *The Elephant and the Tiger: The Full Story of the Vietnam War* (New York: Hippocrene Books, 1990), 396.

41. Hammel, 304. The *NVA* was beginning a withdrawal from Hue, but was leaving substantial forces to fight a rear-guard action. Nolan, 171.

42. Hammel, 336, 340, 347. Nolan, 175.

43. Hammel, 283, 354. Nolan, 101–102, 183–185. Oberdorfer, 230–233. Communist death squads were hard at work alongside the retreating *NVA*. Civilian corpses bore execution-style gunshot wounds and disinterred graves showed many victims were buried alive.

44. Nolan, 184. The ferocity of the battle is reflected in the number of men awarded the nation's highest honor for their actions. Marine Sergeant Alfredo Gonzalez was awarded the Medal of Honor for his gallantry at Hue. This award was also awarded to Army Chief Warrant Officer (CWO) Frederick E. Ferguson, Staff Sergeant (SSG) Clifford Chester Sims (posthumously), and SSG Joe Ronnie Hooper who fought to cut off the *NVA* lines of communication.

Chapter 3

Rock the Casbah: Beirut, 1984

After the 1948 Arab-Israeli conflict, Lebanon became home to more than 110,000 Palestinian refugees. From bases established in southern Lebanon, Palestinian groups, most notably the *Palestine Liberation Organization (PLO)*, conducted raids into northern Israel and bombarded towns with artillery and *Katyusha* rockets. The situation was exacerbated after the Jordanian Civil War from 1971 to 1973, when a large number of Palestinian fighters and refugees fled into Lebanon. Palestinian refugees in Lebanon totaled over 300,000 by 1975, and became essentially a state within a state. Israel responded to the frequent attacks by bombing camps and conducting raids to disperse hostile forces along its northern border. The United Nations and the United States frequently brokered cease-fires, but they were seldom held by either side, and sporadic but deadly attacks recurred.[1]

The situation in Lebanon continued to fester over the years as the *PLO* strengthened in number, established more training centers, and escalated attacks on northern Israel. By 1982, full-time *PLO* military personnel numbered around 15,000 in Lebanon, although only 6,000 were deployed in the south. These forces were equipped with 60 aging tanks, many of which were no longer mobile, and about 250 artillery pieces. Whatever its limitations in conventional war, Israel saw the *PLO* as a potential threat to its northern region and a destabilizing force in Lebanese politics. The *PLO* bases were also believed to be staging points for international terrorism. As the violence escalated, Israel's tolerance reached its breaking point.[2]

In the first half of 1982, attacks by the *PLO* against targets in Israel and abroad escalated. These events invariably triggered an Israeli response, most often in the form of bombing *PLO* camps. The Israeli government came under increasing internal pressure to end these attacks and the deadly cycle of ineffective retaliation. Four times the Israeli Defense Force (IDF) massed an invasion force near its northern border with Lebanon, but each time aborted a ground strike. Israeli patience finally ended with the 3 June 1982 *PLO* assassination attempt on their ambassador to England. The Israeli cabinet quickly met and approved sending ground troops into Lebanon.[3]

This offensive, named Operation Peace for Galilee, was planned as a limited incursion of up to 40 kilometers (25 miles) into Lebanon to just short of the city of Beirut. The goal was to push the *PLO* out of southern

Lebanon and create a security zone adequate enough to place northern Israel out of artillery and rocket range; however, Defense Minister Ariel Sharon and IDF Chief of Staff Rafael Eitan had other plans. They envisioned the complete removal of *PLO* forces from Lebanon. They hoped to eradicate the *PLO*'s military, political, and economic hold over Lebanon. Sharon and Eitan also sought to establish a friendly Lebanese government and strengthen the Lebanese army, which would then maintain security in southern Lebanon. Syrian forces were expected to actively support the *PLO*, but both Israel and Syria seemed determined to isolate any fighting in Lebanon and to avoid an all-out war. The ambitious timeline established by Sharon and Eitan envisioned reaching Beirut within 96 hours. Several members of the Israeli cabinet, suspicious of such an escalation, closely monitored the operation and reserved the right to subject the conduct of the campaign to its approval.[4] (See Map 14.)

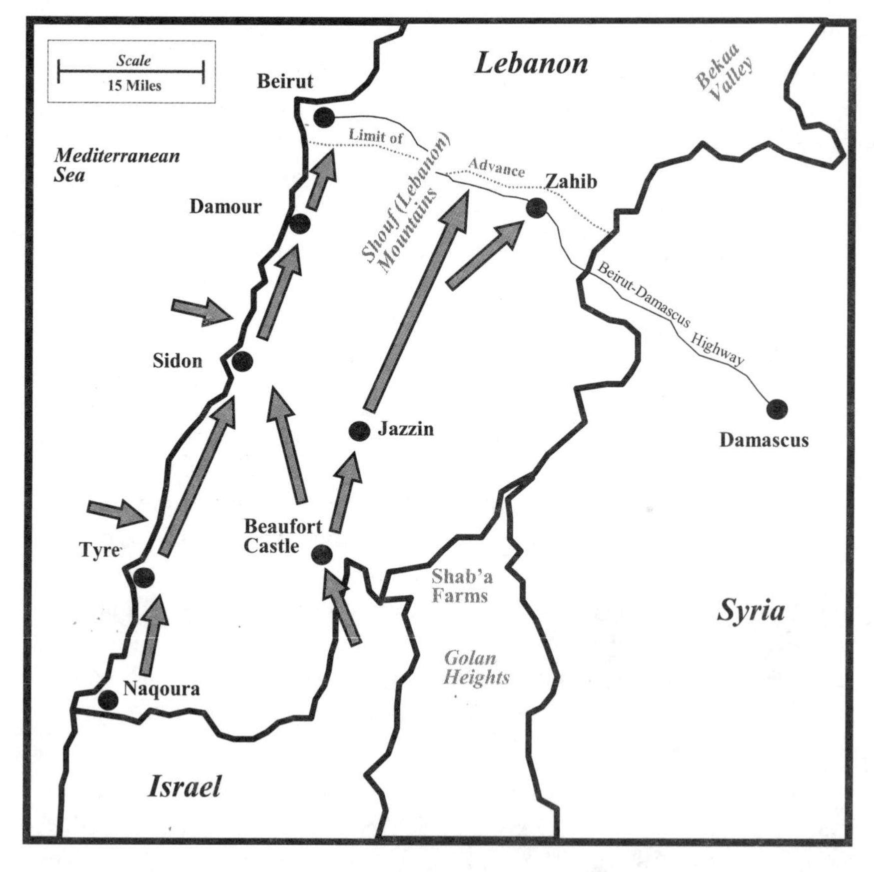

Map 14. The Israeli plan.

The Israeli Defense Forces

This war would be a challenge for the IDF and a direct challenge to its doctrine and history. In past wars, the IDF had fought mainly in the deserts and open country where the strengths of its mobility and long-range firepower were prominent. In Lebanon, it would have to fight in rugged mountains and in confined and congested urban centers. Nevertheless, the IDF had experience in urban operations, most notably in Suez City during the 1973 War. This operation, however, was to confront a major city with over a million inhabitants. Allocating 75,000 troops, 1,250 tanks, and 1,500 armored personnel carriers organized into four independent divisions, an amphibious brigade, a two-division corps, and a reserve division was a formidable task.[5]

Israeli doctrine for urban warfare was typical for the day, calling for the encirclement and bypassing of cities when possible. If necessary, tanks supported by infantry would lead assaults into urban areas. If that proved too difficult, the infantry would lead the tanks. Artillery provided indirect fire support and, in some cases, entered the city to conduct direct fire on stubborn targets. The historic Israeli emphasis on armored units premeditated these tactics, which resulted in a shortage of infantry in their order of battle. In practice, regular armor and infantry units received urban warfare training, but the large number of reservists generally gained only limited urban training in their refresher courses.[6]

For the urban fight, armor units by doctrine were task organized under infantry command as a supporting arm. Command and control and combined arms were emphasized by placing the commanders with forward observers and air liaison officers. Tactical officers at all levels were expected to exercise a substantial level of independence and discretion, a hallmark of Israeli warfighting. Standard Israeli rules of engagement allowed for the application of heavy ordnance to buildings housing hostile forces, but the need to minimize civilian casualties was stressed.[7]

Although artillery and tanks were relied on for their heavy firepower, the IDF also planned to use nonlethal means to secure urban areas. As the situation permitted, the Israelis hoped to use loudspeakers and leaflets to urge civilians to leave the battle area. Friendly civilians were to be employed as guides through the narrow streets or to provide information on enemy forces.

The IDF was equipped with modern, almost state of the art, equipment for that era. The infantrymen carried either the M16A1 rifle or the Galil assault rifle, which were very effective at close range. Additionally for the close-in fight, the infantrymen received hand grenades, radios, grenade launchers, light antitank rockets, and ballistic vests. Heavy support weapons included the M163 Vulcan antiaircraft gun and various models of the 155mm self-propelled artillery piece. The Vulcan, mounted on an M113 armored personnel carrier (APC), and its 20mm cannon proved effective with its ability for high angle and 360-degree fire. Engineer units were equipped with the Caterpillar D-9 bulldozer to clear obstacles and barricades and to create alternate routes. The two primary models of main battle tanks used in the operation were the American-made M60 and the indigenous Merkava (Chariot), although some British-made Centurians and captured Soviet models were still in service. A weakness in the IDF armor was the shortage of night vision optics. The limited night operations relied on illumination rounds fired from artillery and mortars and the tank-mounted searchlights.[8]

The M60 tank series had seen action in limited numbers in the IDF during the 1973 War in both the hills of the Golan Heights and the deserts of the Sinai. A Continental V-12 750-horsepower engine of proven and reliable design powered the M60. The cast hull and turret were of conventional design and layout. The four-man crew consisted of the driver in the hull and the commander, gunner, and loader in the turret. Its main armament was a 105mm rifled cannon, with a 7.62mm coaxial machine gun and a .50-caliber heavy machine gun mounted in the commander's cupola. The Israelis had heavily modified the M60s, improving their fire control and installing Blazer reactive armor. In many cases, the commander's cupola was removed or replaced with a superior one. In its final form the Israelis referred to the vehicle as the Magach 6.[9]

The Merkava, designed and built to Israeli needs after the 1973 War, was the other principal Israeli tank used in the campaign. The first vehicles of the series entered IDF service in 1979. The displacement of armor and locating the 900 horsepower engine to the front were done for maximum crew protection. The main armament of this variant was the venerable 105mm gun, and there were three machine guns as secondary armament. The basic load for the main gun was 85 rounds. The Merkava was unique in design in that the rear doors opened to allow access to a series of ammunition racks holding an additional 200 rounds to facilitate rapid resupply. By removing these racks, the Merkava could carry a ten-man infantry squad under full armor protection. Many contemporary analysts considered the Merkava one of the best tanks in the world.[10]

The dominant number of armored personnel carriers was the American-made M113. Although a reliable vehicle, the aluminum armor made it extremely vulnerable to tank fire and the countless number of RPG-7 rocket grenades held by the Palestinians. Some M113s were retrofitted with additional armor, but not enough to compensate. Not wanting to expose the infantrymen to such danger, the Israelis refrained from committing M113s to close-in fighting, instead using them as transports to ferry men and supplies just short of the battle area and to evacuate the wounded.[11]

The Israeli navy's primary mission was to block the coast of Lebanon and prevent resupply of *PLO* forces. The Israelis had a number of small crafts and a few submarines available, but the Reshef-class patrol boats were the backbone of the effort. These crafts carried six Gabriel missile launchers and two 76mm guns and could operate for extended periods. Another mission given to the Israeli navy was to support amphibious landings and to carry on a deception that more landings were planned with the hope of diverting *PLO* forces sent to guard the coastline. The third mission was to support the ground offensive with naval gunfire.[12]

The Israeli air force was considered one of the world's finest at the time, featuring modern aircraft and highly proficient pilots. The Israelis allocated missions to specific types of aircraft to maximize their capabilities and minimize their limitations and vulnerabilities. For instance, the F-15 Eagles and F-16 Falcons generally provided aerial cover while the F-4 Phantoms, A-4 Skyhawks, French-built Mirages, and Israeli-made Kfirs conducted close air support missions. Ordnance included smart munitions, cluster bombs, missiles, and unguided rockets. Because Arab air defense weapons were historically ineffective, Israeli fighter-bomber pilots were usually able to casually drop their ordnance at 3,000 to 4,000 feet. The Arab antiaircraft guns were generally effective against helicopters, so the Israelis primarily used helicopters to transport supplies and evacuate wounded.[13]

The Christian inhabitants of southern Lebanon deeply resented the *PLO* conversion of their region to a battlefield and formed militias to secure their local areas. The militias in southern Lebanon numbered approximately 23,000 regular fighters organized in small detachments. They were armed and equipped with a myriad of arms and equipment, including antiquated tanks, APCs, and some artillery. Often the Israelis provided this materiel and the training to use it. Most of the militias remained neutral in the campaign and played no significant role. The Israelis hoped the Phalange militia forces in east Beirut, under the command of Bashir Gemayel, would cooperate around Beirut, but their hopes were misplaced. Gemayel's Lebanese Forces (LF), some 8,000 fighters, were organized into companies and battalions and employed by platoons and squads. They were armed primarily with M16A1 and AK-47 rifles, but did have a handful of T-55 tanks, rockets, and artillery pieces.[14]

The Palestine Liberation Organization

In 1982, the 20,000-man *PLO* fighting force was loosely organized and poorly trained by Western standards. In the refugee camps and within the large concentrations of Palestinians in the urban centers, three brigade-size units formed and were often supplied with Soviet-made and Syrian supplied weaponry. The *PLO,* in theory, was ideally suited for urban fighting as its forces were not capable of standing up to the superior Israeli firepower and mobility in open terrain. Factions within the *PLO* prevented any coordinated effort, even at the tactical level. In reality, organization and leadership above the squad level was very poor. Events showed that while under attack the major refugee camps improvised more effective attacks than the *PLO* created during the war. Those forces fought fiercely and tenaciously with limited regard for their own lives. Indeed, many *PLO* forces thought the best they could hope for was an honorable death against the heavy armor of the IDF.[15]

The *PLO*'s two primary weapons were the AK-47 assault rifle and the RPG-7 rocket-propelled grenade. Three-to-six-man squads formed around the large number of RPG-7s available. Covered by supporting fires from the rifles and machine guns, the RPG-7s ambushed and destroyed as many Israeli vehicles as possible in the hope of inflicting heavy casualties. Additionally, there were a large number of hand grenades and mines available, although the *PLO* made little use of the latter. In the streets of southern Lebanon, the roving bands of *PLO* fighters were a deadly menace to any attacker. The Palestinian forces had heavy weapons, including some antiquated and often immobile tanks. They also used the Soviet-made ZPU 14.5mm-heavy machine guns and ZU-23 23mm automatic cannons mounted on light commercial trucks. Highly mobile and effective against soft targets and infantry, these vehicles would suddenly appear, fire, and scatter to safety under the cover of RPGs and small arms fire.[16]

Syrian Forces

Syria had an estimated 30,000 occupation troops in Lebanon under an Arab League mandate issued following the 1975–76 civil war. These troops were in the form of six divisions deployed mainly in the Bekaa Valley and along the main highway between Damascus and Beirut. There were some 600 tanks, mostly the older T-55s and T-62s, and just over 300 artillery and antitank guns. The most formidable units were the *1st Armored Division* with two tank brigades and one infantry brigade and the *1st Mechanized Infantry Division* organized with two infantry brigades and one tank brigade. The *91st Tank Brigade* was also on hand. Syrian doctrine mirrored that of the Soviets, favoring combat in open terrain and avoiding protracted fighting in the cities. Training and equipment of their ground forces were considered somewhat low, and the Israeli invasion caught them by surprise. A major weakness was the Syrian tendency to commit their forces piecemeal by brigade.[17]

The Syrian air force committed over 500 planes to operations in Lebanon. Most of these aircraft were the venerable MiG-21s, MiG-23s, and Sukhoi-22 fighters. The air force's primary mission was to act as an air defense umbrella over the ground forces in the Bekaa Valley, augmented by large concentrations of air defense missile systems. A formidable array of SA-2, SA-3, and SA-6 antiaircraft missile batteries was clustered around the Syrian forces in the valley.[18]

When comparing the forces, the IDF clearly possessed numerical and technological superiority over its foe. This is particularly true as there were few Syrian forces within 40 kilometers of the border. With many of the militia groups remaining neutral in the fight, only the scattered *PLO* forces presented an obstacle to the advance. Few Israeli leaders and planners saw any difficulty in being able to reach the stated objectives.[19]

The First Phase

Israel committed five IDF divisions and two reinforced brigades to the offensive that was planned along three axes of attack. These divisions were aimed against the three major *PLO* concentrations defended by 1,500 to 2,500 fighters. The main effort was in the west and called for two infantry-heavy divisions to converge on Tyre and drive north along the coast to Sidon to link up with an amphibious assault force and drive toward Beirut. Supporting this endeavor was the 162d Armored Division, which was assigned to the center sector to secure the Beaufort Castle, a *PLO* command center held by a brigade-size element, and then proceed to the northwest to Sidon to support the units operating along the coast. To the east, the 252d Armored Division and the 90th Reserve Division, supported by an airborne brigade and an infantry brigade, were to drive northward to destroy the *PLO* forces, push back any Syrian units, and cut the key Beirut-Damascus Highway to prevent further reinforcement or intervention from Syria. If successful, it was hoped the main Syrian forces in the Bekaa Valley would be forced to withdraw from Lebanon.[20]

On the morning of 6 June 1982, IDF began their advance into southern Lebanon. As expected, the *PLO* was the only group to resist the IDF advance. Although many of their leaders fled, the Palestinian fighters proved tenacious. When overrun and scattered, many of them went to refugee camps or into the hills to do battle as small guerrilla units. Already facing international rebuke for launching the invasion, Israel also faced another challenge with the possibility of the small number of United Nations (UN) peacekeepers in Lebanon attempting to block the routes of advance. The IDF was under orders not to engage UN soldiers, and the UN soldiers were instructed to maintain their positions unless their safety was imperiled. This was a dangerous situation for both sides, and any confrontation could have very serious international repercussions. As it turned out, most UN contingents gave the Israeli columns wide berth. However, a Nepalese detachment set up a roadblock on the Khardala Bridge over the Litani River and other UN troops tried to block the coastal road. In each case, the Israeli tanks or bulldozers simply plowed ahead. Fortunately, neither side fired at the other and bloodshed was avoided, but these incidents caused short delays.[21]

The first day of the offensive went fairly well for the Israelis as their columns rolled steadily north into Lebanon. In the east, the Israelis pushed into the hilly country. In the center the air forces heavily bombed the town of Nabatiyah to soften up the defenses for the approaching column. Along the coast, the armored drive closed in on the port of Tyre as a heliborne assault landed troops and even some light tanks as far north as the Zahrani River, some 30 miles north of the border. *PLO* fighters took to ambushing the IDF columns with mines and RPG-7s, but were unable to halt them. Nevertheless, they did cause further delays and casualties.[22] (See Map 15.)

The initial phase of Operation Peace for Galilee was conventional in concept and execution and fitted well with Israeli's equipment and doctrine. The plan to combine armored forces driving deep into enemy territory with support from massed firepower from artillery and aircraft was a replay of many past battles and campaigns. The stated goal remained to reach the 40-kilometer line of advance and close the escape and reinforcement routes of the *PLO* as quickly as possible. In this, the IDF was in its element and the disparity between it and its enemy was clearly apparent as the *PLO* lacked mobility and firepower. But like so many wars in history, easy victory can be elusive because the enemy has a say in the matter.[23]

The first major obstacle in this rapid advance to the final objective was the city of Tyre and the sprawling refugee camps surrounding it. The IDF had prepared for the worst in urban combat in this sector and had given units refresher training and additional equipment. To minimize civilian casualties, the Israelis dropped leaflets instructing the residents to avoid ground and air attacks by assembling on the beaches. The Israelis employed all branches of its services in this well-planned and executed operation. An amphibious landing to the north of the city blocked the main

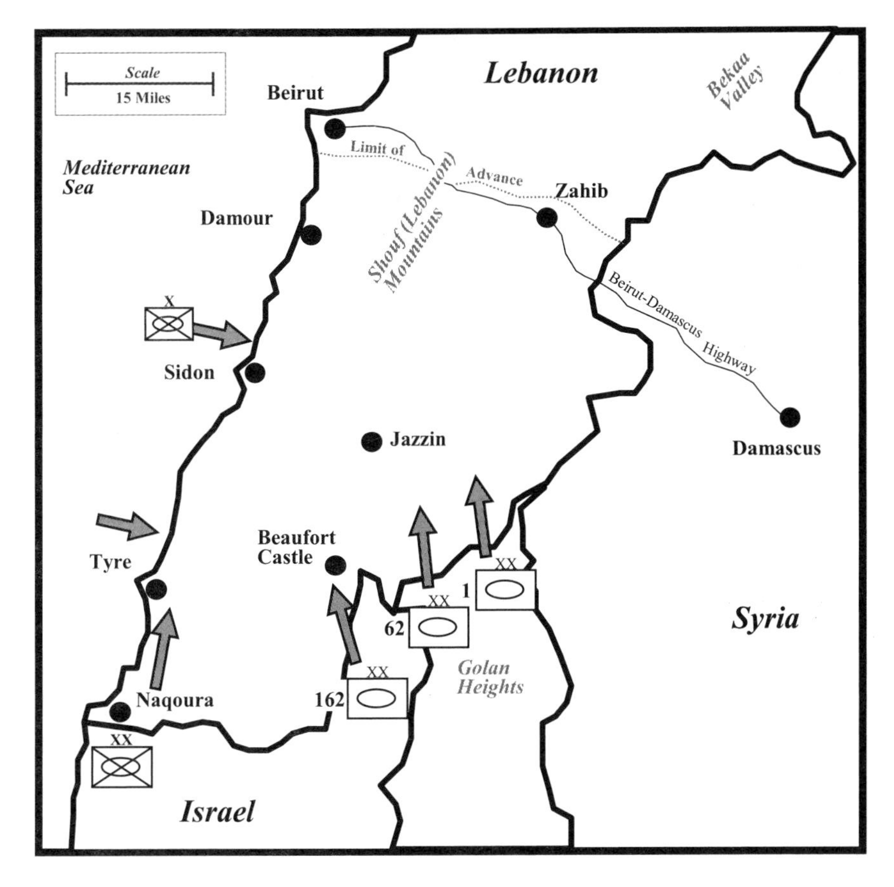

Map 15. 6 June 1982.

route of egress or reinforcement while naval gunfire rendered support. The city was essentially isolated and under seige.[24]

After artillery and air support missions blasted the camps and *PLO* positions in the city, the IDF ground forces then entered the camps with tanks in the lead and APCs close behind. The tents and ramshackle buildings of the refugee camps presented a maze to the attacker and a multitude of firing points for the numerous RPG-7 gunners. Beyond the camps, the city of Tyre was formidable with its substantial construction. The coming house-to-house fight appeared to be the tank crewman's nightmare. Much to the surprise of the senior Palestinian leadership and the Israelis themselves, the battle for Tyre was bitter but not the bloodbath the armor units feared. Tanks fired point-blank at street-level bunkers, while the paratroopers and infantry engaged upper level positions with small arms and mortars. The speed and shock of the assault dazed the

Palestinian leadership who were unable to coordinate an effective defense. Unexpectedly, the heaviest fighting was over in a day, although it took another four or five days to clear Tyre completely of resistance. Here the high explosive antitank (HEAT) and sabot rounds often failed to penetrate the concrete structures in the city. In response, the IDF employed 155mm self-propelled guns in a direct-fire mode. These proved very effective in reducing strongpoints and in some cases collapsed entire buildings.[25]

The Israeli timetable was now out of synch because of the battle for Tyre, but the armored thrust up the coast continued. The city of Sidon was similar to Tyre for the attackers, including a brigade amphibious assault, but as the southern headquarters for the *PLO* it was expected to have more determined defenders. When the IDF tanks and infantry stormed into the camps and city, the light resistance astonished them. The Israelis did not know it at the time, but the Palestinians had largely abandoned the city and withdrawn north into Beirut. Determined *PLO* fighters, however, continued to conduct sporadic hit-and-run attacks until the very end of the conflict.[26]

The Israeli columns in the center and the west were tank-heavy and faced sporadic resistance that generally was far less than expected; even so, it was not a simple walkover. Ambushes were common on the tortuous dirt roads in the hills and in the villages. At times, the infantry had to dismount and lead the tanks through the streets, thus slowing the speed of the advance. As the IDF pushed northward, it encountered increasing numbers of Syrian troops. Sporadic skirmishes became pitched battles. To bypass a particularly strong Syrian position, the IDF engineers gouged a 20-kilometer track in the rugged hills to allow an armored task force to proceed. Massed armor and firepower dominated these fights, and Israeli armored units continued to push to the north. The net effect of these battles with the Palestinians and the Syrians was to throw off the ambitious timetable in the center and eastern sectors.[27]

On 7 June, the Israelis conducted a rare night attack on the Beaufort Castle in the center sector. The castle was an imposing structure rising some 700 meters from the Litani River Valley and dominated the terrain for miles. For years, the *PLO* had used it as a headquarters and for directing artillery and rocket fire into northern Israel. Defending this redoubt were 1,500 *PLO* fighters armed with a wide assortment of light and heavy weapons supported by some artillery and a few tanks. Using the headlights of APCs and illumination rounds, the Golani Brigade took the position by storm and, although there were heavy casualties, eliminated this troublesome obstacle. Although successful here, a fight near the town of

Ayn Zhaltah was another matter. On 8 June, the Syrians were able to stop the Israelis using heavy artillery fire and antitank rockets and missiles. This action prevented the 162d Armored Division from cutting the Beirut-Damascus Highway in this sector and outflanking the Syrian defenses in the Bekaa Valley.[28] (See Map 16.)

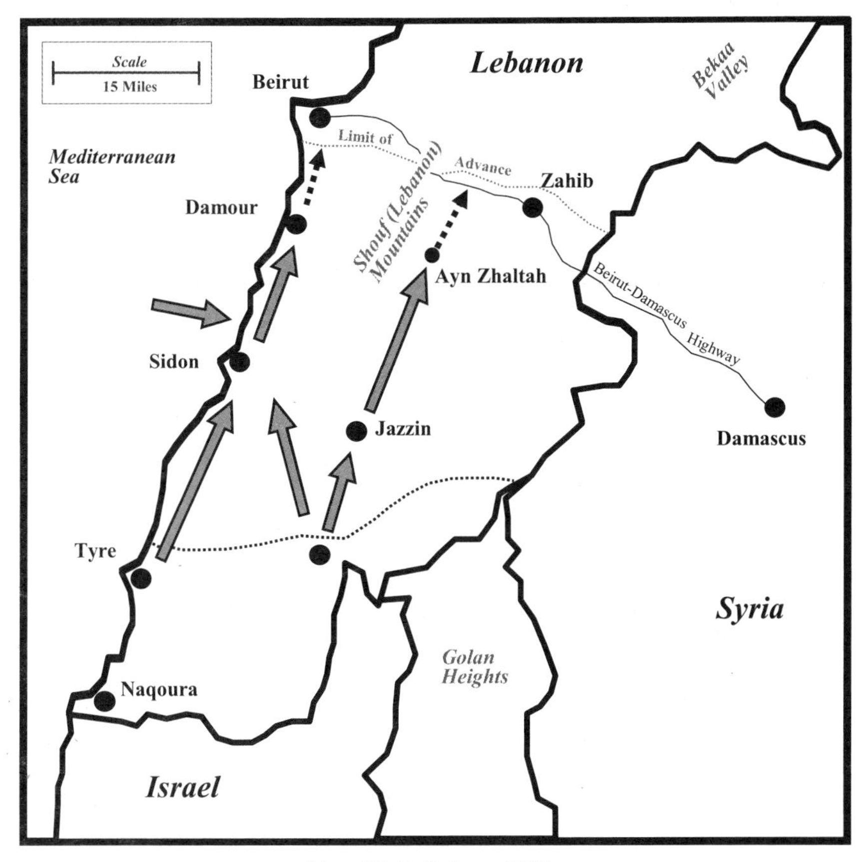

Map 16. 7–8 June 1982.

A key event occurred on 9 June with the decimation of the Syrian air force and the virtual elimination of the surface-to-air missile sites that were deployed to the Bekaa Valley two years earlier. More than 90 Israeli fighters engaged initially 60 of their Syrian counterparts in supersonic dogfights. Meanwhile, two massive air strikes hit the Syrian missile sites and armor units in the valley. The number of planes reported lost vary greatly between both sides, but apparently nearly 80 Syrian planes were destroyed while Israel lost none. With this action, Israel won air supremacy over southern Lebanon. Except for the threat of the man-portable SA-7 missiles and antiaircraft guns, Israel could conduct interdiction and ground support missions with virtual impunity, as long as the aircraft remained at altitude beyond the effectiveness of these weapons. With control of the skies guaranteed, the IDF launched a strong attack against the Syrian *1st Armored Division*, which was positioned south of the Beirut-Damascus Highway to protect this vital supply link to the estimated 30,000 Syrian soldiers in Lebanon. Massive air strikes pounded the Syrians, and two Israeli divisions launched a frontal attack on the entrenched positions. These coordinated attacks severely mauled the Syrian division, but casualties were heavy for the Israelis too. Their attack in this sector halted just short of severing the highway. The Syrians hurriedly sent the *3d Armored Division,* equipped with a number of T-72 tanks, into Lebanon to help stem the Israeli tide.[29] (See Map 17.)

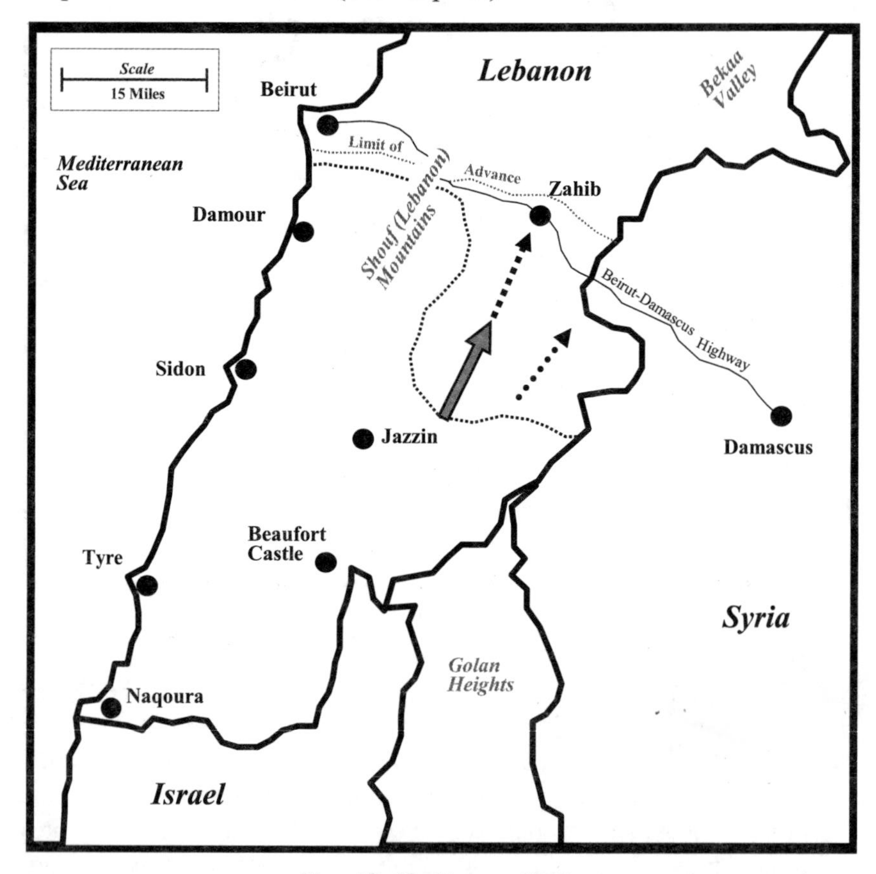

Map 17. 10–11 June 1982.

Although stalled in the east just short of the Beirut-Damascus Highway, the Israelis had made steady if not rapid progress along the coast. After a brief fight in Damour, the IDF pushed on toward Beirut. As the Israelis got closer to the capital, the *PLO* resistance became more efficient and determined. By 13 June, the IDF entered the fringes of west Beirut and made contact with the Phalange militia under Bashir Gemayel. The Israelis had hoped the Lebanese militia would enter the streets of west Beirut and clear the *PLO* forces from the city; instead, it remained neutral in the fight. If the IDF wanted to take Beirut, it would have to do so on its own. Although such an operation needed cabinet approval as the city lay beyond the stated goals of the offensive, Sharon told the IDF to occupy and strengthen its positions around west Beirut.[30]

Even though the timetable was not followed and there were heavier casualties than originally planned, the Israelis were close to meeting the stated objectives of the offensive. The *PLO* in southern Lebanon was shattered. The Syrians were contained, yet fighting, in the Bekaa Valley. Facing a humiliating defeat, the Syrians asked the Soviet Union for direct intervention. The Soviet Union rejected that option, but accelerated the delivery of weapons, equipment, and advisors. For Israel, the delay and casualties meant political trouble. Sharon came under increasing pressure to answer to the public on the war effort and to justify its cost. At the same time, he did whatever he could to gain cabinet approval to extend the campaign's objectives to include the encirclement of Beirut. His argument was that if the *PLO* was simply pushed out of southern Lebanon, it was free to return once Israeli forces were withdrawn. Sharon believed a lasting peace was possible only if the Palestinians were driven out completely. *PLO* Chairman Yassar Arafat did not find this good news and expressed his desire for a cease-fire, if only to gain time to negotiate or strengthen his military position. He knew Israel lacked the will to accept the heavy casualties associated with urban warfare and that time and attrition would work to his advantage. Naturally, the international media would expose the suffering of the civilian populace, bringing further pressure on Israel. Under urging from the United States, Israel relented to a temporary cease-fire, but Arafat rejected calls for removal of the *PLO* from Beirut and Lebanon. Frustrated with the stalemate, Sharon ordered the IDF to move into Beirut.[31]

Battle of Beirut

Once considered the "Paris of the Middle East," by 1982 Beirut was but a shell of its former glory. Gone were the days when the city attracted tourist and businessmen from Europe and Asia who frequented the market places and beaches. The Lebanese Civil War, begun in 1975, was primarily responsible for the state of economic depression. Political and cultural division resulted in the Christians controlling the eastern half while the Muslims and *PLO* occupied the city in the west. A narrow growth of trees and bushes stretching for ten miles through the heart of Beirut, called the Green Line, physically manifested this division. This terrain feature had three crossing points and served as a stark and real divide between the city's east and west halves. West Beirut was about ten square miles in size and had seen the most visible damage over the years, leaving few buildings intact. Electricity, water, and municipal services were sporadic; food and fuel were in short supply for the 600,000 residents.[32]

The *PLO*, occupying the southwest quarter of the city, had turned west Beirut into a Palestinian capital in exile. Its headquarters was established in the Fakhani district where a few buildings rose to 14 stories, but the construction was generally of lower quality than that along the prime beachfront property. The headquarters building had been modified for eventual hostilities with the addition of three underground levels. Nearby was a sports stadium that was converted into a major ammunition depot and a recruiting and training center. Bunkers and logistics caches dotted the city. Streets in this district were often too narrow for large military vehicles like tanks. Located at the southern edge of the city were the Beirut International Airport and several refugee camps. The terrain was flat and sandy, and the camps, housing about 200,000 Palestinian refugees, could pose a very serious problem. (See Map 18.)

The Palestinian fighters planned to concentrate on protecting the *PLO* headquarters and the three refugee camps of Sabra, Shatilla, and Burj al-Barajinah; but the Palestinians were short of heavy weapons. There were approximately 40 T-34 tanks, 30 DM-2 scout cars, 70 obsolete antiaircraft guns, and 24 BM-21 rocket launchers. The 16,000 Palestinians vowing to fight to the death would shoulder the bulk of any fighting. Armed with AK-47 rifles or RPG-7s and holed up in thousands of apartments in west Beirut, they had taken advantage of the short-lived cease-fire by frantically fortifying their positions, mining the approaches to the city, and emplacing booby traps.[33]

The Palestinians had fewer than 2,000 Lebanese Muslim militiamen as allies. They were the leftist groups Sunni Murabitun and Shiite Amal. The Murabitun defended the port area and National Museum crossing, while Amal concentrated its forces on protecting the Shiite slum areas. Additionally, Syria had posted the *85th Mechanized Brigade* composed of some 2,300 men equipped with about 40 T-54/55 tanks and APCs, two artillery battalions, and a battery of 57mm antiaircraft guns. This brigade had suffered heavy damage in the fighting south of Beirut and the nearly

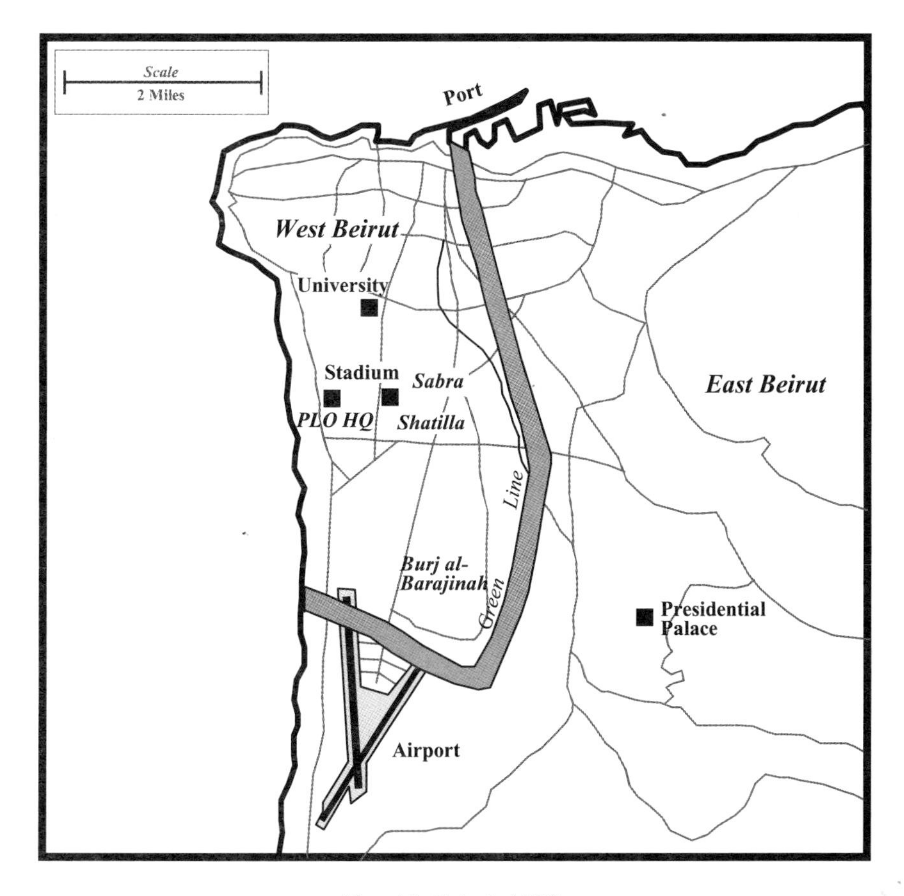

Map 18. Beirut, 1982.

constant bombing by the Israeli air force. The Syrians were deployed to the southern parts of west Beirut, in the relatively open areas that were best suited for their armor, and they had a position near the Soviet embassy. The Syrians also controlled many of the Palestinian and militia forces within their grasp. The splinter groups within the *PLO*, the militias, and the Syrians acting in their own interests formed a command and control nightmare. Each group would fight its own battle, in its own way, for its own purpose.[34]

The Israelis knew of the *PLO*'s command and control problems and other woes, but the sheer number of combatants and the urban setting gave them pause. Fortunately, the terrain gave the Israelis two distinct advantages. First, there was a series of mountains ringing the south and east of Beirut rising to over 6,000 feet. Artillery and heavy weapons had an excellent view of the city below. Second, the bulk of the Palestinian fighters were deployed to cover the camps, which were located on more open terrain away from the city proper. The IDF could thus concentrate its bombing on the *Fakhani* district with the *PLO* headquarters and the three refugee camps while minimizing the risk to the majority of the Lebanese inhabitants. (See Map 19.)

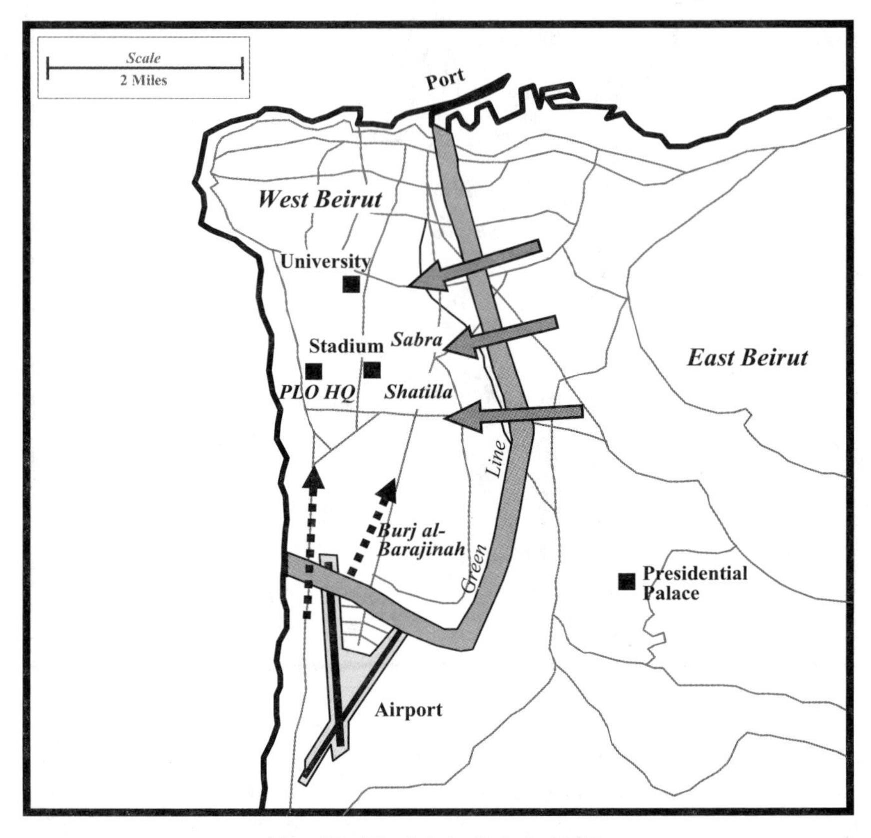

Map 19. Attack into Beirut, 1982.

With the *PLO* rejecting the call to leave Lebanon, Sharon set out to force the issue. To do that the IDF had to occupy the dominating terrain around Beirut and cut the strategic Beirut-Damascus Highway. The Syrian brigade, who put up a stubborn resistance, occupied those heights. For 13 days, the IDF and Syrians battled for their possession, but by 26 June Israel controlled the high ground and 13 miles of the highway. Also, even though the Lebanese Forces under Bashir Gemayel were officially neutral, they firmly held the north and east sectors of Beirut. Under these conditions, the bulk of the *PLO* forces and the Syrian *85th Brigade* were trapped in the capital. To keep up the pressure, the Israelis periodically lobbed artillery shells into the Palestinian camps and city districts. Sharon renewed his demand that the *PLO* leave Lebanon and offered the Syrians clear passage to leave with their weapons and equipment. The Lebanese army would enter west Beirut to disarm the *PLO*. These demands and offers were rejected, and Israel continued to tighten its grip around Beirut militarily and through an economic blockade. On 1 July, Israeli aircraft thundered over the city in mock bombing runs. Leaflets and radio broadcasts urged civilians to leave the city before the coming battle. The battle for Beirut was to begin in earnest.[35]

On 3 July, an IDF column of armor and infantry advanced toward the Burj al-Barajinah refugee camp in the southern part of the city. This force managed to gain a shallow penetration into the camp after heavy fighting. The success of this action was limited, but the Israeli government publicly announced its willingness to continue the siege through the winter if necessary.

The attack on the Burj al-Barajinah camp did have an effect on the Israeli use of armor in urban warfare for the rest of the siege. Unlike in the cities along the coast, here the Palestinians had put up a stiff resistance and the RPG-7s had proven effective against the Israeli tanks. With increasing losses in tanks and crews, the IDF became reluctant to use tanks in another massed assault. Believing time was on their side, the Israelis settled on using periodic artillery strikes while continuing to tighten the blockade around the city. This trend continued for the next two weeks, then escalated after the *PLO* launched several raids into Israeli positions. The response on 21 July was a massive aerial bombardment augmented by naval gunfire, heavy artillery, and tank fire. These attacks were maintained until the end of the month.

Faced with growing discontent in the cabinet, Sharon was more determined than ever to force a resolution. In the early morning of 1 August, an Israeli task force of tanks, infantry, and paratroopers attacked and took control of the Beirut International Airport. During the next three days, air and artillery incessantly pounded west Beirut. This was a prelude to something much bigger.[36]

Into the City

The war's largest ground operation against Beirut commenced on 4 August under massive air bombardments, artillery shelling, and heavy fire along the shoreline. Collateral damage was significant, including the American University Hospital, the prime minister's building, the Central Bank, the offices of Newsweek and United Press International, and the two luxury hotels. Residential areas also experienced widespread damage.

From east Beirut, Israeli Defense Forces crossed into west Beirut at the three checkpoints on the Green Line with the main effort aimed in the direction of the *PLO* headquarters. Engineers and bulldozers led the way for the tanks, infantry, and paratroopers. Bitter street fighting occurred and the Israelis managed to capture the National Museum and the Hippodrome; nevertheless, they failed to break through to the *PLO* headquarters.[37]

The supporting attacks were launched from the south of Beirut, capitalizing on the area near the airport seized earlier. One thrust proceeded up the coast for just half a mile before being stopped by heavy *PLO* resistance. The other attack began at the airport and struck toward the northeast with the aim of driving a wedge between the refugee camps. The camps had been largely abandoned, but the heavy fire from *PLO* machine guns, RPG-7s, and artillery stopped these supporting attacks.

Although the attacks were halted short of their objectives, the Israelis demonstrated they were willing to commit ground forces to the systematic destruction of the Palestinian forces in an urban environment. Pressed militarily and economically, and increasingly isolated on the diplomatic front, time was running out for Arafat and the *PLO*. Arafat finally agreed to withdraw from Beirut with several conditions. To maintain pressure on the Palestinian leadership, on 12 August Sharon ordered the heaviest bombardment of Beirut to date. The aerial and artillery assault lasted for 12 hours and focused on the *PLO* headquarters and the refugee camps. It is estimated that over 130 people were killed and over 400 were wounded, mostly civilians. The heavy civilian casualties shocked the Israeli people, and this unilateral action by Sharon prompted the cabinet to strip him of all military authority. As a result, all future military operations needed the approval of the cabinet or prime minister. On 19 August, the cabinet approved the *PLO* evacuation plan and under the watchful eye and protection of a UN multinational force, the *PLO* began its departure on 21 August. The last of the 14,000 Palestinian fighters in Lebanon left on 3 September. The siege of Beirut was over.[38]

The total IDF losses for Operation Peace for Galilee were 344 killed and over 2,000 wounded, almost half of which were sustained during the fight for Beirut. The *PLO* lost over 4,500 in Beirut, and the various Lebanese militias and the Syrians lost about 3,500. The civilians of the city fared poorly with figures estimating 6,000 killed and wounded.[39]

The direct result of the Israeli offensive was the withdrawal of most of the Palestinian military forces from Lebanon and their establishment in Tunisia. This ended the most serious threat of attacks to northern Israel, but tensions remained high with Syria and their closely allied government of Lebanon. The conflict proved disastrous for Syria militarily, losing large numbers of armored vehicles and aircraft, but it retained control of over a third of Lebanon. An accord was reached in 1983 to set conditions for an Israeli withdrawal, but it was never enacted and peace and stability remained elusive in the troubled region. Nonetheless, Israel began a phased withdrawal from Lebanon in 1985, leaving behind a number of Lebanese militias, known collectively as the South Lebanese Army, to operate a security zone. The last IDF units withdrew from this zone in 2000.[40]

In Retrospect

The IDF launched Operation Peace for Galilee with the expectation of fighting in an urban environment and had prepared by issuing specialized equipment and conducting training. It had an established doctrine and experience from previous wars and from previous security operations within Israel and the occupied territories. The casualty-adverse IDF was fully aware that armor attacks in urban areas without sufficient infantry support were costly and task organized to compensate for the tank-heavy force structure. In urban fights, tanks and tank units were generally placed under infantry command, which was quite a departure from the norm.

The Israelis lost relatively few tanks in the operation considering the widespread use of man-portable antitank weapons used by the *PLO*. It is not clear whether this was due to the poor marksmanship of the RPG-7 operators or the inability of the weapon to cause major damage to the vehicles. It was probably a combination of both. Measured in a controlled environment against a flat surface, the RPG-7 was capable of penetrating ten inches of rolled homogeneous armor. By design, the M60 and Merkava tanks have sloped sides, and the Israelis had modified many of their vehicles with reactive armor. Many IDF tanks sustained multiple hits and continued to fight, but the Merkava's performance was outstanding. It proved to be the safest tank in action, as no single crewman was killed in the operation. In addition, the add-on reactive armor on the M60s and Centurians proved their worth in protecting their crews.[41]

Intended to fight in the open and not in urban areas, both the M60 and Merkava tanks were of contemporary design. These tanks were unable to maneuver down many of the narrow roads and alleys, and the long cannon barrels were extremely restricted in their ability to traverse. Their machine guns lacked sufficient elevation to provide suppressive fire or engage targets in the upper stories of buildings. Palestinian snipers forced the Israeli tank commanders to abandon temporarily their habit of directing their fire and units from an open hatch. In spite of their shortcomings in design, the

Israeli tanks served well by providing direct fire support to their accompanying infantry and stunning the *PLO* defenders with the speed and shock of their attacks.

As expected, supporting vehicles such as the M113 APC proved especially vulnerable in the cities as a hit from an RPG-7 meant certain casualties. Designed solely to carry infantrymen to the battlefield where they would fight dismounted, buttoning up the hatches made the situation worse as the crew and infantry could not fire their weapons. The IDF infantrymen quickly learned to dismount and fight on foot. The Merkava tanks that had their ammunition racks removed compensated somewhat for the lack of a capable infantry carrier, but they could not carry all of the supporting infantry and were further limited in the number of rounds they could shoot. Some Merkavas served as makeshift ambulances, but this practice removed the powerful weapon from the front line. Another makeshift solution was to employ armored engineer vehicles to carry infantry, with a corresponding loss of engineer capability.

The IDF made considerable use of smoke in the battle for Tyre and Sidon, but used it sparingly in the siege of Beirut. It had proved useful in preventing effective targeting by RPG-7 gunners, but apparently this caused more problems than it prevented. The smoke often interfered with the hand and arm signals used by small unit leaders and blinded the tank drivers, thus slowing the pace of an advance or attack. To compensate for the lack of smoke, the Israelis used mortars for suppressive fires. Mortars were favored for their psychological effect and high angle of fire that allowed their use in built-up areas. However, the 60mm and the 81mm weapons, common in infantry formations, could not penetrate the upper roofs of the modern buildings. Conversely, the heavier Soviet-made 120mm in the hands of the Syrians and Palestinians could penetrate Israeli-held buildings with ease.[42]

All combatants used their antiaircraft guns in ground support roles as well. They all had a sufficient elevation and traverse to target upper stories of buildings and a high rate of fire to suppress enemy forces. Particularly fearsome was the Vulcan 20mm cannon, which was able to penetrate most buildings with a rate of fire of over 2,000 rounds per minute. The Syrians employed the venerable ZU-23 gun of 23mm, and the Palestinians had a few of these as well.

In summary, the IDF proved adept at using tanks and armor in an urban environment. Equipped with capable armored vehicles manned by highly trained crews and led by capable leaders, they produced decisive results in the fights for Tyre and Sidon. Faced with the more daunting challenge of the major city of Beirut, they performed well but suffered heavy damage and casualties inherent in such warfare and against an enemy in possession of effective weapons and the will to fight. In this case, overwhelming firepower often made up for the shortcomings in organizations and in vehicle design. If this operation had any lasting effect on Israeli doctrine remains to be seen, as there has been no subsequent operation of such size and scope. Since they did not radically changed their order of battle after the war, it appears the Israelis were generally satisfied with their doctrine. Happening during the Cold War and preparing to fight on the German plains, the United States did not fully scrutinize its own doctrine using the lessons of the fight for Beirut.[43]

Notes

1. Chaim Herzog, *The Arab-Israeli Wars: War and Peace in the Middle East* (New York: Vintage Books, 1984), 339–340. The term *Katyusha* rockets includes a wide range of unguided rockets, generally of Soviet origin. The most common is the 122mm version, often associated with the BM-21 rocket launcher. The maximum range of most weapons was about 45 kilometers. Not known for their accuracy, these were purely area weapons.

2. Kenneth M. Pollack, *Arabs at War: Military Effectiveness, 1948–1991* (Lincoln, NE: University of Nebraska Press, 2002), 524. See also Yazid Sayigh, *Arab Military Industry: Capability, Performance, and Impact* (London: Brassey's Defense Publishers, 1992), 524.

3. Herzog, 342.

4. Ibid., 345.

5. David Eshel, *Chariots of the Desert: The Story of the Israeli Armoured Corps* (London: Brassey's Defense Publishers, 1989), 162. Anthony H. Cordesman and Patrick Baetjer, *The Lessons of Modern War, Volume I: The Arab-Israeli Conflicts, 1973–1989* (Boulder, CO: Westview Press, 1990), 165–166, 169. Few would rate the operation in Suez City an Israeli success.

6. Eshel, 156. George W. Gawrych, "The Siege of Beirut," in *Block by Block: The Challenges of Urban Operations,* ed. William G. Robertson (Fort Leavenworth, KS: U.S. Army Command and General Staff College Press, 2003), 213–214. This is an excellent source for further reading.

7. Gawrych, 214. It was observed that concern about civilian casualties and property damage naturally declined as casualties rose.

8. Pollack, 525. Cordesman, 171, 173, 184. Some sources name the M60 tank series the Patton. Although technically correct, few people called the M60 the Patton, reserving that name for the M48 series. The Israelis did have a number of M48s in service, but these were not used in large numbers in this operation.

9. Cordesman, 87–88, 173. R.P. Hunnicutt, *Patton: A History of the American Main Battle Tank* (Novato, CA: Presidio Press, 1984), 210, 225. Author's note: The M60A2 was an engineering disaster. Designed to fire the 152mm Shillelagh missile, this variant never overcame "teething" problems and was eventually scrapped. In the author's opinion, the original M60 cupola was abysmal. The vision blocks were poor, the ammunition storage was small, and the M85 machine gun was difficult to operate and prone to jamming. The small hand cranks were inadequate to traverse and elevate the machine gun quickly. The common trait of all M60 tank commanders was the scar across the knuckles that resulted from clearing jams during gunnery qualifications.

10. Eshel, 157–161. Cordesman, 172–173. The Merkava was simply a rolling arsenal.

11. Cordesman, 171, 174–175.

12. Cordesman, 216. Gawrych, 225.

13. Cordesman, 193, 195–196.

14. Herzog, 339.

15. Cordesman, 119–121. The three *PLO* brigades were called *Karameh*, *Yarmuk*, and *Castel*.

16. Cordesman, 119, 171, 183. The US Army in Somalia in the 1990s often called the trucks mounting heavy weapons "technical vehicles."

17. Pollack, 523–524. Herzog, 344, 346. Syria mechanized many of its infantry organizations and increased the number and mobility of its air defenses after the disastrous 1973 War. It also greatly expanded its commando forces, but stripped many of the best personnel from its infantry forces to do so. Cordesman, 122, 173. The new T-72 tanks were a special concern to the Israelis as there was some doubt whether the 105mm gun would penetrate the armor. As it turned out, it did.

18. Herzog, 344, 347. Cordesman 194–195.

19. Pollack, 525.

20. Eshel, 162–163. Pollack, 525–527. Although the main effort was along the coast, the heaviest and most powerful units were deployed to the east to confront the more capable Syrians.

21. Cordesman, 136–137, 149.

22. Pollack, 528. Cordesman, 137.

23. Herzog, 345.

24. Cordesman, 138–139.

25. Eshel, 163. Cordesman, 139. Gawrych, 226. The 155mm SP guns were later used in Beirut in the direct fire mode.

26. Cordesman, 139, 141–142.

27. Eshel, 163–164.

28. Herzog, 343, 346. Pollack, 530–531. This was one of the most important factors in preventing the complete destruction of the Syrian army in Lebanon.

29. Eshel, 166. Herzog, 347–348. Pollack, 538–540. Cordesman, 200–201. The losses represented over 15 percent of the Syrian air force and many of its best pilots. The air battles did divert most of the Israeli air force from the critical ground support mission and was one factor in the failure to cut the highway at this point. See also Pollack, 532–534, 536–537. The Syrian *1st Armored Division* lost over 150 tanks, and an entire brigade was destroyed. This was the combat debut of the vaunted T-72 tank. It did not prove to be as good as some analysts had predicted.

30. Herzog, 349–350. Pollack, 541–542. Some analysts, such as Dr. Gawrych (214–215), contend that the IDF could have taken Beirut "on the run" at this point with a good chance of success. They contend the *PLO* was stunned and shocked by the speed and power of the Israeli advance and could not offer much as a defense. On 13 June Jerusalem announced the death of Major General Yekutiel Adam, the highest-ranking Israeli officer ever killed in battle. Up to this point, Israel had suffered 214 killed, 1,176 wounded, and 23 missing in action.

31. Herzog, 35–37. Gawrych, 217. Cordesman, 148–149. Sharon's objectives were clearly open-ended and might have dragged Israel into an all-out war with Syria that it was not prepared to fight and with no predictable end. If the IDF were able to plan for these final objectives at the outset and driven faster

and harder in southern Lebanon, it is likely Israel would have achieved its major military goals by 20 June.

32. Gawrych, 209–210.

33. Cordesman, 144, 151. Gawrych, 210–211, 219.

34. Cordesman, 120–121. Pollack, 543–545. Pollack asserts that the Syrian armor forces in particular were inept at combined arms warfare and incapable of maneuvering and adjusting artillery. Reconnaissance was almost nonexistent, and soldiers and crews were not disposed to clean and maintain their weapons and equipment. He marvels that the Syrians were not destroyed outright and faults the Israelis "start and stop" pace of the offensive for this. See Steven J. Zaloga, *T-54, T-55, T-62* (New Territories, Hong Kong: Concord Publishing, 1992) for technical details of this older Soviet armor.

35. Gawrych, 219–221. The IDF and LF established checkpoints that stopped all traffic into western Beirut. Only medical personnel, police, and firemen could pass. Most of the Syrian forces managed to slip out of western Beirut before the noose was fully tightened.

36. Dissent against the war emerged throughout Israel and even within the ranks of the IDF. Gawrych, 228. In late July a brigade commander, Colonel Eli Geva, refused an order to fire his artillery into areas of western Beirut arguing such bombardment would cause numerous civilian casualties. He was relieved of his command, but such dissent within the military was unprecedented in the annals of the IDF and a shock to society.

37. Herzog, 352. Gawrych, 221.

38. Gawrych, 222–223. Herzog, 352.

39. Cordesman, 152. Gawrych, 223.

40. Herzog, 352–354. Saad Haddad was the founder and commander of the South Lebanon Army, and had commanded a militia battalion during Operation Peace for Galilee. Israel armed and supplied his forces. He died of cancer in late 1984.

41. Eshel, 167. Cordesman, 155–156. The *PLO* was essentially an occupation force, neither trained nor equipped for the task at hand. If it had fought as well as the Syrian commandos, for instance, the Israeli losses would have been far worse.

42. Eshel, 167. Israeli tank commanders had traditionally stood upright in their turrets during battle to facilitate rapid target acquisition and command and control.

43. Cordesman, 221. Cordesman contends that many analysts in the West learned the wrong lessons from the war, focusing almost exclusively on the technological aspects and not on the employment of the weapons themselves.

Chapter 4

Headlong into Hell: Grozny, 1995

Russia's invasion of Chechnya in January 1995 ranks among the worst military engagements of the 20th century. The fighting centered on the city of Grozny where a hastily assembled and unprepared Russian force squared off against a Chechen force of regulars and guerrillas equipped with Russian weapons and a belief in their cause. Although the Chechens held their own for three weeks, they eventually lost the city to the Russians. The city changed hands again in 1996 and yet again in 2000. Each Russian offensive used tanks and armor in the urban fight. Consequently, the battles for Grozny had a deep impact on the major world powers' opinions of using armor in cities.

The battles for Grozny are an important example of large-scale operations using armor in urban combat. For many contemporary analysts, the battles for Grozny represented the future of modern warfare. In Grozny, a technologically advanced army battled for control of a large city held by a small irregular force. High casualties, massive collateral damage, and heavy losses in vehicles and equipment all point to the apparent folly of using tanks in the urban fight. This chapter will discuss the 1995 battle for Grozny and examine why the Russian armor did not achieve the desired results.

For students of Russian history, the conflict was no surprise as the region of Chechnya has long suffered. Chechnya, located in southeastern Russia in the Caspian Sea region, was a major producer of oil. Fiercely independent, the people of Chechnya traditionally resisted all authority by those they consider foreigners and paid the price often by enduring military occupations and pogroms. Perhaps the most brutal was in 1944 when Joseph Stalin deported the entire population to Central Asia. Thousands died as a result and this nurtured an increased long-standing hatred toward Russia. Soviet ruler Nikita Khrushchev allowed the people of Chechnya to return to their homeland 13 years later, but their attitude toward Russia remained unchanged. Over the years, the Chechens waited impatiently for a chance to gain their independence.[1] (See Map 20.)

Taking advantage of Russian political chaos, Chechen President Jokhar Dudayev declared the republic's independence in October 1991. Friction and competition between Russian President Boris Yeltsin, Soviet President Mikhail Gorbachev, and the Russian Supreme Soviet prevented a swift reaction to the breakaway republic. Half-hearted economic sanctions and political isolation were imposed, but there was no significant military

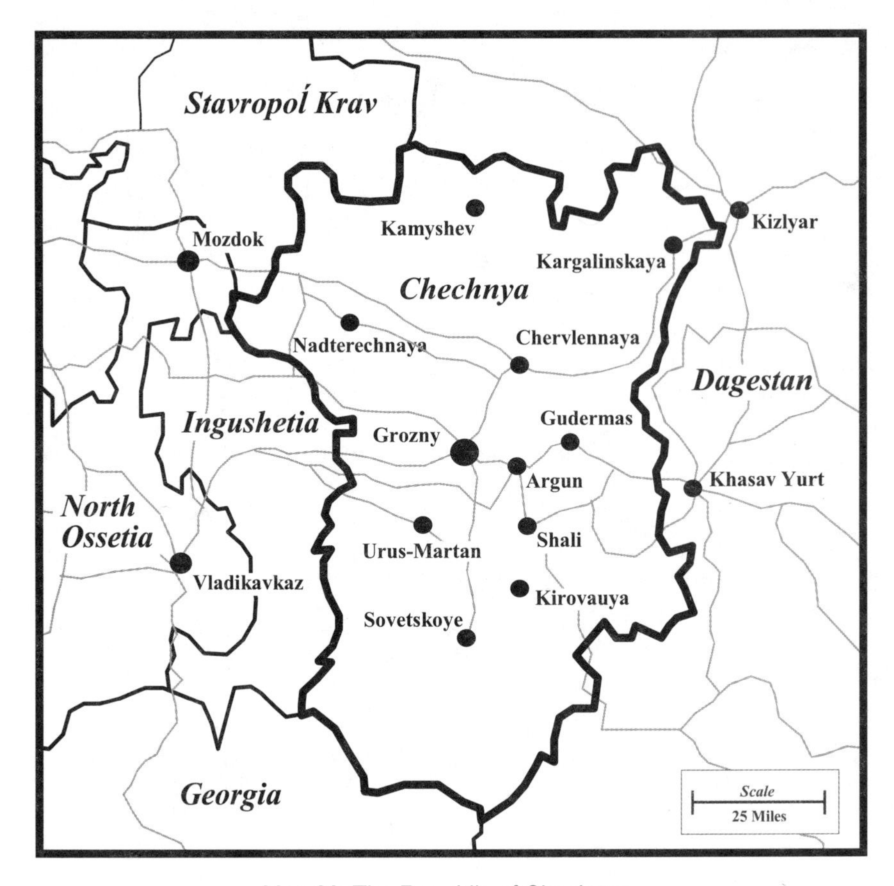

Map 20. The Republic of Chechnya.

action. Pro-Russian factions were provided with money and weapons, but they failed to bring Chechnya back into the Russian fold. Humiliated with the repeated failures, in December 1994 President Yeltsin used his newly consolidated power to call for a full-scale military intervention to reestablish Russian dominion over the republic by disarming the "criminal elements in control." A confrontation was inevitable, as Chechens remained defiant and continued toward their goal of full autonomy.[2]

The Russians made plans to invade Chechnya to reestablish control and to set an example for other republics that entertained similar notions. The plan had four major components and was not particularly complicated. The first phase was to isolate Grozny by sealing the border and deploying a combination of army and interior ministry troops to form a cordon around the city from the north, west, and east. A route from the city to the south was to remain open to allow Chechen forces to leave the city. This opening phase was expected to take about three days.[3]

The second phase of this operation was committing powerful armored forces into the city itself, with the objective of rapidly securing the Presidential Palace and government buildings. Other objectives included radio and television stations and the utilities that provided water, electricity, and sewage removal. The Russian planners envisioned this phase to take approximately four days and expected the Chechen rebels to avail themselves of the opportunity to flee the massive armored columns by using the aforementioned open corridor.[4]

Phase three would be the pursuit of the Chechen forces into the mountains and establishing a government friendly to Russia. This was to take about ten days. Eliminating small pockets of resistance that held out in the mountains, a process that was expected to take a few weeks or months, was the fourth and final phase. The Russians expected light resistance throughout the operation due to their overwhelming firepower capability.[5]

Grozny was the capital of the Chechen Republic and in 1994 had a population of nearly 490,000 people. Being a major industrial and petroleum center, Grozny had many multistory buildings and covered some 100 square miles. It was truly a modern city with electricity, a sewer system, and the other needed infrastructure for a city of its size. In the 1980s, it was widely regarded as one of the most beautiful cities in the region, with modern architecture, plazas, parks, and wide thoroughfares. The people of Grozny, and indeed the region, were fiercely independent and prepared to give any invader a hostile reception. This was a society where the clan was dominant and possessed a strict moral code and high motivation. This was the great strength of the Chechen fighters—each knew why they were fighting. They were prepared to give their lives for a cause they believed in and hoped to take many Russian lives with them in the process.[6]

Russian Order of Battle and Planning

It is extremely difficult to ascertain the Russian order of battle during the 1994–95 conflict in Chechnya. Most units were of a composite nature, and manning was usually far below the norm. A rough estimate is that the original force used in Chechnya was composed of 19,000 from the Russian army and 4,700 from the Ministry of the Interior. There were approximately 80 tanks, 208 armored infantry vehicles, and over 180 artillery pieces. Reinforcements were sent into the republic in the following weeks and months until the total number of troops was about 58,000. Large numbers of aircraft were committed to the invasion, coming mostly from the 4th Air Army stationed in the North Caucasian Military District. Although impressive on paper, from the start there were problems with coordinating the troops from the different ministries. The Ministries of

Defense, Internal Affairs, and Internal Security also committed forces to the operation; however, they did not integrate their efforts to achieve a common goal.[7]

The Russians deployed the T-80 main battle tank for the intervention in Chechnya. In production since the late 1970s, the T-80 was a fully modern vehicle and, at the time, the most advanced tank variant in Russia's arsenal. Its main armament was a 125mm main gun capable of firing a variety of antitank and antipersonnel rounds. Secondary armament included a 7.62mm coaxial machine gun and a 12.7mm heavy machine gun in the commander's cupola. The T-80's armor was considered excellent and was often further augmented by explosive reactive armor blocks. A three-man crew was possible due to the use of an automatic loader, and when issued, thermal sights allowed the crew to engage targets in limited light. The Russians also used an earlier version of this tank, the T-72. The T-72 had many of the same features but its armor was not as effective and it used a diesel engine instead of a gas turbine. Most Western analysts rated these tanks as among the world's best.[8]

The BTR-80 APC was the standard infantry vehicle for the campaign in Chechnya and Grozny. This eight-wheeled APC had a crew of three and could carry seven infantrymen, and the 14.5mm heavy machine gun and 7.62mm coaxial gun provided direct fire support. The BTR-80 was a refinement over the BTR-60/70 series, with improvements in the power plant, gun elevation, and crew egress being the primary differences. The older variants did see service in Chechnya as well. The other infantry-fighting vehicle to see service was the tracked BMP series. Fielded in the early 1980s as an improved version of the BMP-1, the BMP-2 was the most numerous model. The main armament was the 30mm automatic gun and the AT-5 *Spandrel* antitank missile launcher. With a crew of three, the BMP-2 could carry seven infantrymen. All BMP variants had relatively thin armor, and due to the compactness of the vehicle, any round penetrating it resulted in a personnel, mobility, or firepower kill.[9]

On 29 November, Yeltsin ordered Chechnya to disarm and surrender to Russian rule. Infuriated by their refusal, he gave the military two weeks to prepare for the invasion of Chechnya. The haste in planning the operation was evident, and in the process the Russians made some very questionable assumptions. First was the assumption that the Chechens would not resist the overwhelming firepower of such a large force. Another assumption was that the military was as capable and ready as it was during the Cold War era. It was not. Training, discipline, logistics, and equipment readiness rates were low in comparison to that bygone era. Units were often composite organizations, with some ad hoc battalions made up of elements from five to seven different formations. Often crews hardly knew one another. Many soldiers went into battle with only rudimentary training with their weapons and no knowledge of fighting in urban terrain.[10]

If there were some bad planning assumptions made, the Russians also made some serious mistakes and omissions prior to the campaign. Perhaps foremost was the treatment of communications. Due to a shortage and unreliability of encryption devices, the decision was made to transmit all messages in the clear. This allowed the Chechens to monitor the Russian communications and to insert bogus traffic over the net. There was no preparation in the use of relays or antennas to overcome the interference of high-rise buildings and power lines common in an urban environment. Additionally, the Russians did not have enough maps for the tactical commanders, and the standard 1:100,000 scale maps were not adequate for the urban fight.[11] Detailed intelligence was also missing during the planning and preparation phases. The situation inside Grozny was almost unknown, and there is little evidence the Russians had fully analyzed the urban terrain or surrounding areas. There was little or no knowledge of Grozny's underground sewer and tram systems, as well as the back alleys and streets where enemy forces were bound to be waiting. Another key factor in the coming fight was that Russia's military doctrine was based on an offensive across northern Europe and against North Atlantic Treaty Organization (NATO) forces. The intervention in Chechnya called for a far more flexible operation and specialty training. The Russians were woefully unprepared, and many field officers questioned their units' ability to carry out such an operation.[12]

Chechen Order of Battle and Planning

Similarly, the Chechen order of battle is virtually impossible to calculate because these forces were loosely organized and fluctuated in strength. Propaganda on both sides inflated the numbers. Apparently, there were 15,000 to 35,000 armed Chechens in the republic, many of whom received military training in the old Soviet army. The typical Chechen fighting group was a three-or-four-man cell, and three or four cells would often combine temporarily against a specific target. Probably the largest group was the *Abkhasian Battalion* led by Shamil Basayev, one of President Dudayev's most trusted associates. Basayev's battle-hardened force consisted of some 500 men who had fought in Abkhazia against Georgians in 1992–93. This force occasionally moved in a group as large as 200 while conducting an ambush or staging a raid. A number of foreign fighters also rallied to the cause. What these forces lacked in organization and training they more than made up for in tenacity and improvisation, including mobilizing entire villages to engage and harass the Russian armored columns.[13]

For heavy weapons, the Chechens had about 12 to 15 working tanks of the T-54 and T-62 models. These tanks, inferior to the Russian T-72s and T-80s, were used primarily as pillboxes in Grozny to cover key intersections and facilities. There were also approximately 40 BTR and BMP series vehicles in the city and about 30 artillery pieces scattered about. The Chechen air force was destroyed on the ground in the first hours of the war and was not a factor in the fight.[14]

If the Chechens lacked heavy weaponry, they had an abundance of light arms and rocket propelled grenades. The large inventory of weapons and the stockpiles of ammunition were the result of the chaotic withdrawal of Russian forces from the republic two years earlier. Essentially, the Russians had armed and supplied their enemy well. The AK-47 assault rifle was the basic arm of the Chechens, but the powerhouse in their arsenal was the RPG-7 rocket propelled grenade launcher. A skilled user could operate the RPG-7 like a mortar to shoot over buildings, as an area weapon against troop formations, or as a precision weapon when fired directly at armored vehicles.[15]

The Chechens employed a novel approach in preparing their defenses in Grozny. Instead of the traditional use of strongpoints, Dudayev and his allies decided to focus almost exclusively on mobile hit-and-run tactics. Using their intimate knowledge of the city, each district leader sent out small teams seeking targets of opportunity and preparing ambushes. The plan was to let the Russian forces move into the city, then surround and isolate individual units. Antitank weapons would attack the tanks and infantry vehicles in quick hit-and-run actions. An extensive number of booby traps were employed to make operations by Russian infantry extremely hazardous. The Chechens positioned the few tanks and armored vehicles they possessed to cover main avenues of approach or as bait to draw attacking Russians into kill zones.[16]

Another tactic the Chechens planned to use was to pose as friendly civilians only to lead Russian patrols or convoys into ambush. Many of the Chechens spoke Russian and donned Russian uniforms for a number of clandestine and deceptive actions. Facing a massive amount of artillery, the Chechens planned to "hug" Russian units, remaining so close that supporting fires would be limited or equally deadly to both sides. The Chechens also planned to vacate positions quickly before massed fires fell, and capitalize on the public relations fiasco when schools, hospitals, and churches were leveled in the process. Mobility was achieved by using civilian vehicles and trucks to transport personnel and logistics. To the Russians, the Chechens would appear as deadly phantoms, difficult to locate and harder to hit. To coordinate all of this activity, the Chechens established an impressive communications network using cellular phones and off-the-shelf radios that also allowed extensive monitoring and manipulation of Russian transmissions.[17]

The Invasion

The Russian intervention in Chechnya began with an air campaign on 20 November 1994. The Russian air force attacked the airfields around Grozny, destroying the Chechen aircraft on the ground. Across the republic airfields, bridges, key roads, and towns were bombed in preparation for the coming ground invasion. With complete domination of the air, Russian planes were able to attack at will.[18]

Yeltsin ordered the ground offensive to commence on 11 December in spite of the deteriorating weather that would hamper movement in the rough terrain. In column formation, three groups made their way into Chechnya with the goal of reaching Grozny in the shortest time possible. The northern group advanced from Mozdok, the western group from Vladikavkaz and Beslan through Ingusjetia, and the eastern group moved in from Dagestan. The columns advanced with the airborne troops in the lead, the other army units following, and the Ministry of the Interior units in the rear. Mi-24 helicopters and SU-25 ground attack aircraft loitered overhead for support, but the winter weather limited their availability and effectiveness.[19] (See Map 21.)

To their surprise, heavy resistance met the Russians. There were no set-piece battles in the open. Instead, the Chechens used their large numbers of antitank weaponry to ambush Russian columns in the forests and hills along the route to Grozny. Favorite targets were the rear echelon and Ministry of the Interior troops. The purpose of these attacks was to delay the Russians in reaching Grozny to give the Chechen fighters time to prepare the defenses in the city. It was there that Dudajev planned to fight the Russians in earnest. The resistance was successful for it was not until the last days of December that the Russian forces reached the outskirts of Grozny. By 31 December, the Russians finally surrounded the city on the west, north, and east. Stepping up the aerial and artillery strikes, the next phase of the operation was at hand. With reinforcements arriving from neighboring districts, the total Russian strength was approximately 38,000 men, 230 tanks, 353 APCs, and 388 artillery pieces. A hastily composed plan dictated the attack would take place along four axes converging on

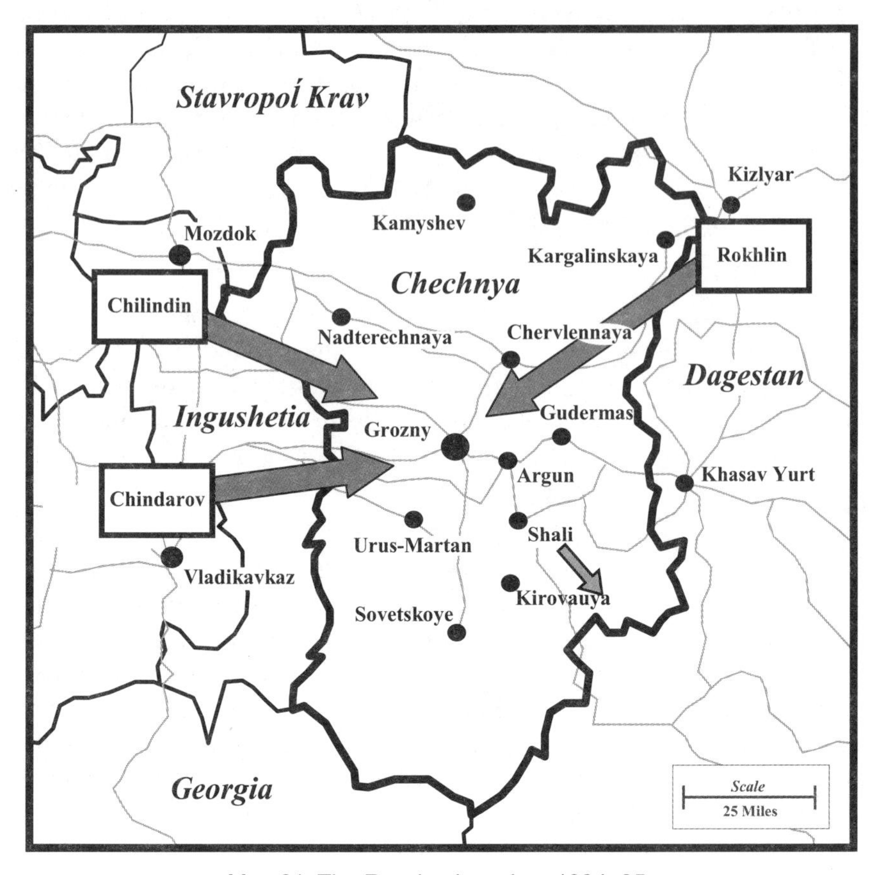

Map 21. The Russian invasion, 1994–95.

the city center. Meanwhile, two *Spetsnaz* groups deployed by helicopters to monitor the Chechen rear areas south of the city.[20]

New Year's Day marked the beginning of the Russian ground assault on the city; immediately, the plan and the attack became a catastrophe. To seize the key points of the city quickly, the Russians committed their armored forces in column hoping the size and speed of the movement would shock the defenders into submission or flight. Dismounted infantry were not used to protect the tanks and vehicles as this would slow the rate of advance. This intended show of strength would prove just the opposite.

The Chechen defensive strategy and tactics worked beyond expectations. The small teams used RPG-7s and grenades to destroy the columns' lead and trail vehicles. Once trapped, the vehicles in the column were attacked and destroyed one by one. The poor communications and lack of urban training doomed the Russian soldiers. All but one of the Russian columns were quickly stopped in their tracks. (See Map 22.)

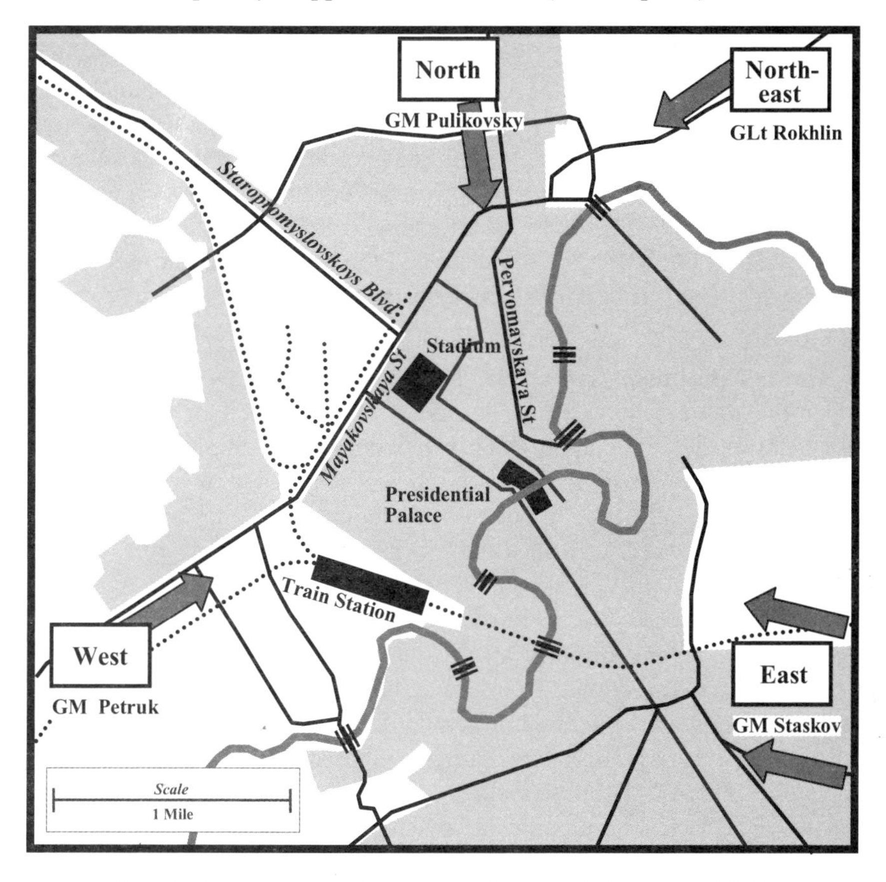

Map 22. The Russian advance into Grozny, 31 December 1995.

Only the northeast group under Lieutenant General Lev Rokhlin reached the city center near the Presidential Palace and the Chechens' headquarters. The 131st Independent Motorized Infantry Brigade (also known as the Maikop Brigade) made it through the city virtually unopposed and seized the railway station. Heartened by their apparent easy success and unaware of the Russian disasters elsewhere, the soldiers parked their tanks and BMPs in the open city center and milled about. Other elements of the brigade remained parked along a side street as a reserve force. After halting the Russian advances in other parts of the city, the Chechens converged on the train station and surrounded it. As before, they destroyed the Russian lead and rear vehicles on the side streets, trapping the remaining armor in the narrow confines. Caught in a shooting gallery, the tanks were unable to respond effectively because their main guns could not depress low enough to engage targets in the basements or high enough for targets on the rooftops. The complete loss of Russian mobility enabled the Chechens to destroy systematically the column from above and below with RPGs and grenades. The brigade commander, Colonel Ivan Savin, called frantically for reinforcements that never came. Over the next two days, the Maikop Brigade lost 20 of its 26 tanks and 102 of its 120 infantry vehicles. Of its 1,000 men, over 800 were killed, including Savin, and 74 were taken prisoner. The remnants of the brigade withdrew by 3 January, leaving behind the bodies of the dead and the hulks of burned-out tanks and vehicles. Total Russian casualties for the first days of battle were well over 2,000.[21]

The horrific losses forced the Russians to withdraw from the city proper and regroup. It was time to assess and modify their tactics. One basic fact was clear: the taking of Grozny would involve fighting for every building and room. To do this the Russian's configured their formations in battalion-size assault detachments with dismounted infantry as their core. Tanks and APCs were still in the fight, but in a supporting role. Tanks were positioned to provide direct fire support against enemy strongholds, to help seal off areas, and to repel counterattacks. While moving, tanks would remain behind infantry at a distance beyond the effective range of enemy antitank weapons. To improve the vehicles' survivability, metal nets and screens were mounted on armored vehicles. Since Russian tanks could not elevate or depress their main gun tubes and coaxial machine guns adequately, ZSU-23-4 self-propelled, multibarreled, antiaircraft machine guns moved forward in columns to engage targets on the rooftops and in basements.[22]

The Russians began their new offensive on 3 January and for the next 20 days the Russians and Chechens battled for the streets. Under cover of massed artillery and aerial bombardment, the Russian ground forces moved forward. The onslaught did not have the desired effect, as the Chechens dug in deeper. Casualties on both sides were high, and the collateral damage to the city was massive. This heavy fire was often counterproductive as the shelling turned the local population against the Russians. The fight illustrated the shortcomings of the initial Russian plan, for instead of retreating through the open corridor to the south, the Chechens were using it to bring reinforcements and logistics into Grozny. With inadequate or nonexistent maps and shoddy communications, Russian units had poor situational awareness. Unit boundaries were unclear and formations often did not know the whereabouts of other friendly units. Fratricide was common, and when a unit advanced it often exposed a flank or was attacked from above or below because the adjoining unit was not in position to support. The Chechens compounded this by targeting and eliminating a large number of Russian radio operators.[23]

The bloody conflict shocked and frustrated the Russian soldiers leaving many on the brink of mutiny. Instead of simply disarming rebellious formations, the Russians found themselves in a war with the total population of Chechnya. Reports from Russian journalists often contradicted the official statements from Moscow, and public support for the endeavor crumbled. Western journalists reported from the scene and international condemnation followed. Russian soldiers, facing an enemy they often could not see and a civilian population who opposed them with equal zeal, became even more demoralized.[24]

As in all urban conflicts, the fighting was three-dimensional and the choice of weaponry and tactics was vital. During the initial Russian thrust into the city, the RPG-7 was the dominant weapon in destroying the armored columns. Although it continued to play a key role for both sides, the sniper eclipsed it in this stage of the war. Both the Russians and Chechens used snipers widely and praised their success, although the Chechens were generally more effective. Snipers were apt to cause panic in the ranks, eliminate leaders and key personnel, stop or slow Russian convoys, or force convoys to different routes altogether. Often those alternate routes led to an ambush. Because many of the Chechens wore civilian clothing, the Russian checkpoints were reduced to searching for snipers by looking for signs of bruising caused by weapon recoil and powder burns on the face and forearms.[25]

An effective weapon the Russians possessed was the man-portable Shmel incendiary rocket launcher. Its fuel-air explosive warhead was devastating in close confines and was often rated on par with a 122mm artillery shell. With a 600-meter range, it was capable of quickly engaging almost any target of opportunity in Grozny. The Russians also made wide use of the Mukha grenade fired from a variant of the RPG-7. The thermobaric warhead for this weapon was deadly efficient against personnel sheltered in confined spaces. Both of these weapons were used as a substitute for supporting artillery, which was difficult to arrange due to poor communications and the Chechen tactic of remaining close to Russian units.[26]

The Chechens continued to execute a defiant and impressive resistance, but could not hold on indefinitely. The onslaught of heavy Russian firepower eventually pushed the Chechens to the south. On the night of 18 January, the Presidential Palace was hit by two bunker-busting bombs, collapsing several floors and forcing Aslan Maskhadov, the tactical commander of

Grozny, and his staff to move the headquarters to a hospital on the south side of the Sunzha River. On 19 January, the Russian forces stormed and took the Presidential Palace. On 20 January, Yeltsin declared the military phase of the operation in Chechnya almost complete and the Ministry of the Interior responsible for establishing law and order.[27] The fighting was not quite over, but on 23 January Russian forces managed to cordon off the southern approaches to the city and cut off the remaining Chechen fighters in Grozny. The Chechens still held the southeast corner of the city, but not for long. As heavy air and artillery ordnance rained down on the city, Shamil Basayev was forced to withdraw most of his men from Grozny. On 7 March, the Russians could finally declare full control over the city.[28] (See Map 23.)

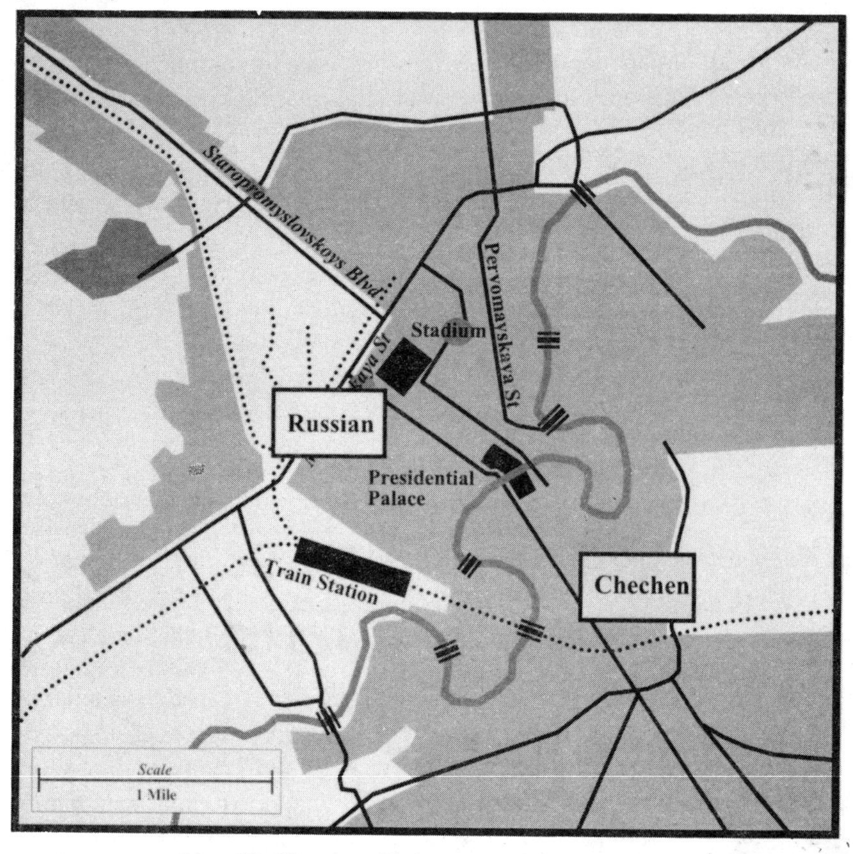

Map 23. Situation, 20 January–13 March 1995.

After Grozny

The battle for Grozny was exceptionally costly, with the civilian population taking the majority of the casualties. The Chechen losses are not known, but the official Russian figures for casualties in the battle of Grozny were 1,376 killed and 408 missing. Over 200 tanks and armored vehicles were lost, as well as the prestige and pride of the Russian armed services. Nevertheless, law and order did not come easily or quickly to Grozny. Terror attacks continued, particularly under the cover of night. The citizenry did not readily accept the installed pro-Russian government. The government operated almost under siege, only surviving due to the presence of the Russian military and by taking refuge in military facilities.[29]

The Russian high command continued with the third and fourth phases of their plan, which were to push the hostile Chechens into the countryside and then beyond the mountains to the south. Hoping to recreate their success in Grozny, the Chechens employed the strategy of fighting to successive cities, such as Shali and Argun, all the way to the base of the mountains. This strategy hoped to negate the Russian advantage of air and ground firepower and to allow the Chechen fighters to blend in with the local population, thus forcing the Russians into the tight urban terrain where it was difficult to distinguish combatants from civilians. Again, the heavy collateral damage incurred usually helped the Chechen fighters secure the support of the local population. By the end of May 1995, after weeks of heavy fighting the Russians controlled about two-thirds of the republic. However, the Russians were unable to crush fully the rebellion. In March 1996, some 2,000 Chechen fighters infiltrated into Grozny and seized large sections of the city. Their intent was not to hold Grozny indefinitely, but to demonstrate that neither the pro-Russian government nor its masters were in control. Facing eroding support at home, Yeltsin offered peace to secure his reelection in the summer. In August 1996, the Russians signed a humiliating cease-fire ending the conflict for the moment.[30] (See Map 24.)

Chechens once again struck Grozny on 6 August 1996, three days before Yeltsins' inauguration. Led by Shamil Basayev, more that 1,500 fighters infiltrated the city, secured key areas, laid seize to the 12,000 Russian troops holed up in garrisons posted throughout the city, and surrounded Russian garrisons in Argun and Gudermes. The Russians immediately reacted to this threat with a massive and indiscriminate application of firepower. Two days later, armored columns were organized to move to the relief of the beleaguered garrisons. In a replay of 1 January 1995, these columns ran headlong into Chechen ambushes and were decimated by RPG-7 fire. On 9 August, talks began between the opposing sides that led to another cease-fire. Later talks produced the Khasavjurt Agreement and the withdrawal of all Russian forces from the republic.[31]

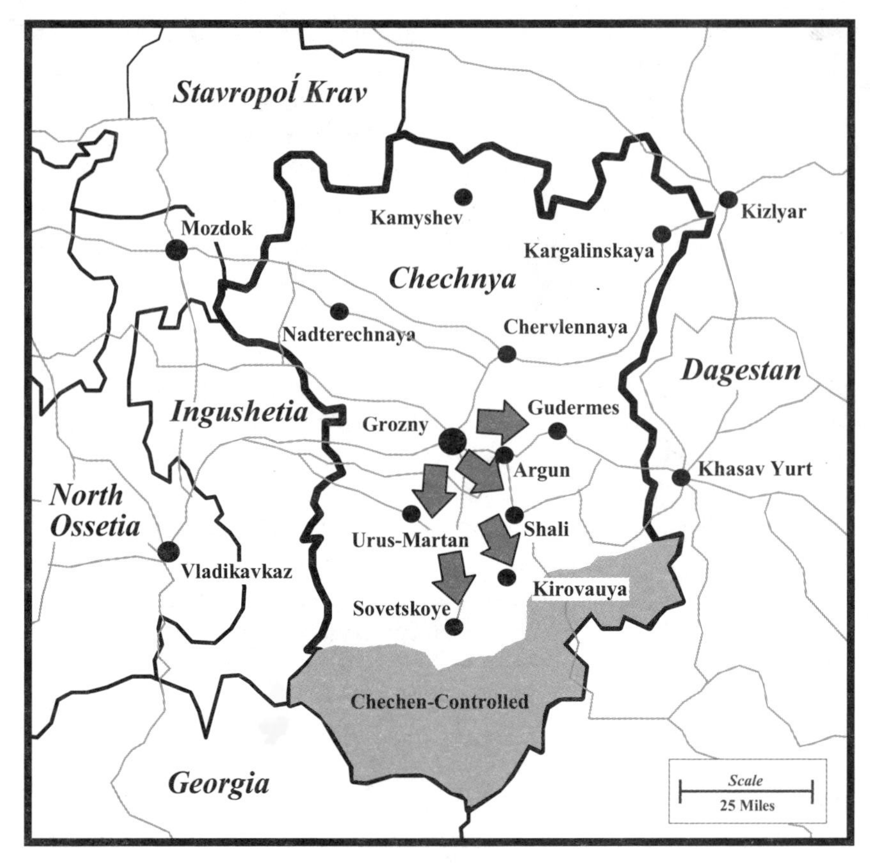

Map 24. The battle of successive cities, 1995.

In Retrospect

The debacle of the Russian intervention in Grozny has been fodder for the analysts and military leaders who have maintained that tanks and armor are not suited for combat in urban areas. In spite of an overwhelming superiority in personnel, weaponry, and firepower, the Russians needed nearly three months to secure Grozny and several months more to make the same claim with the rest of the small republic of Chechnya. The determined Chechen defense and high Russian losses had a substantial impact on military planning for the next decade. Only a bold commander would commit his armor to battle in the streets.

The Russian planning for the operation was obviously inadequate and based on a number of false assumptions. The planners were far too optimistic in assessing the ability and readiness of the Russian military to intimidate the Chechens into disarming or fighting. In giving the Russian military only two weeks to plan an operation of such magnitude,

Yeltsin perhaps doomed the operation from the start. There was simply not enough time for a proper staff analysis and dissemination of information. Russian units had little time to prepare equipment, train, and assemble. Considering the distances many of these units traveled to participate in the opening battles, it is probable their entire time was spent in transit. This was certainly a "come as you are" war. Whatever readiness condition and training a unit had prior to hostilities, that is what it was on commitment, oftentimes less due to mechanical breakdowns.[32]

In 1994, the Russian military was but a shadow of the old Red Army of the Cold War era. There were chronic manpower shortages, and morale was low due to deteriorating living conditions and low and irregular pay. Many formations were simply cadre units. The Russians had a doctrine for urban warfare based on its World War II experience, but there had been little or no training in city fighting for years. Funding shortages inhibited unit training, and there had been no exercises above division-level in two years. There was also a chronic lack of spare parts for vehicles and equipment.[33]

This malaise percolated down and affected all ranks. Many of the poorly trained conscripts had not fired their weapons beyond their basic training. In fact, some soldiers had not even finished their introductory training. Many tank and armored vehicle crews were unfamiliar with their equipment and all were unprepared for the complex combined arms fighting in urban terrain. The composite nature of the units as they were thrown into battle guaranteed that the men had not trained together as a team. As fighting in cities is usually at platoon-level and below, this was a recipe for disaster, especially up against a wily and determined foe. The rank and file of the Russian military was told they were sent into Chechnya to simply disarm illegal formations and establish law and order. It was quite a shock for them to encounter the tenacity and lethality of the Chechens. Morale further plummeted and despondency set in quickly as Russian public opinion turned against the war.[34]

Although lighter than Western designs, most analysts gave high marks to the assortment of infantry fighting vehicles possessed by the Russians. The T-80 and T-72 tanks were assessed to be on par with their Western counterparts; with their advanced armor, they were expected to take a great deal of punishment. But, the battle for Grozny exposed some of the flaws of Russian armor. The tanks could not depress or elevate their main guns adequately to engage targets in basements or high in the buildings or rooftops. The armor protection on the top and rear was relatively thin, and the RPG-7s and antitank mines stood a good chance of penetrating these areas. The long tubes of the 125mm cannon had a very narrow traverse radius in the narrow streets that limited firing generally to the frontal arc. Clearly, the Russian tanks were designed to fight in the open country and not in the confines of a large city. To their credit, many tanks took multiple hits by RPG fire before their destruction. However, when the lead and trail vehicles in the column were disabled, even the best tank and crew found they were stationary targets with little ability to return fire. Much the same can be said for the infantry fighting vehicles, but their armor could not take the same amount of punishment.[35]

With the failure to take Grozny quickly, the Russians returned to their traditional means of taking an urban area. Unconcerned with collateral damage or civilian casualties, they used massed artillery and aerial strikes systematically and literally to pulverize the city into rubble. This application of firepower eventually overcame the difficulties in communications, intelligence, unit coordination, and a host of other problems the Russian units faced.

The lessons of Grozny are sobering for anyone who contemplates using armor in an urban environment. Yet it was not a fair test of the tanks' ability to fight in the city streets. Had the Russians followed the basic tenets of using combined arms with adequate communications and control, the results could have been far different. This fight emphasized the need to train completely the crews and troops in their weapons, tactics, and doctrine to employ them effectively. The battle for Grozny was actually a historical aberration in the use of armor in the urban fight. Far different results were possible had the Russians employed and supported their armor correctly.

Notes

1. Timothy L. Thomas, "The 31 December 1994–8 February 1995 Battle for Grozny," in *Block by Block: The Challenges of Urban Operations,* ed. William G. Robertson (Fort Leavenworth, KS: U.S. Army Command and General Staff College Press, 2003), 161. See also Pontus Siŕen, "The Battle for Grozny," in *Russia and Chechnia: The Permanent Crisis,* ed. Ben Fowkes (New York: St. Martin's Press, Inc., 1998), 90–92.

2. Thomas, 163–164. General Jokhar Dudayev came to power as a result of an uprising in 1991 that expelled the Chechen communist leader Doku Zavgayev. Dudayev was elected in October 1991. He was killed by a Russian missile in 1996. Siŕen, 100–101, 116.

3. Olga Oliker, *Russia's Chechen Wars, 1994–2000: Lessons from Urban Combat* (Santa Monica, CA: RAND Corporation, 2001), 9–10. Thomas, 167.

4. Oliker, 9.

5. Ibid., 10–11.

6. Thomas, 162.

7. Thomas, 165. Thomas has listed some of the various units in this excellent piece on the battles for Grozny. Oliker, 23. Anatol Lieven, *Chechnya: Tombstone of Russian Power* (New Haven, CN: Yale University Press, 1998), 102–103. The acronym MVD stands for Ministerstvo Vnutrennikh Del.

8. Richard Simpkin. *Red Armour: An Examination of the Soviet Mobile Concept* (Washington, DC: Brassey's Defense Publisher, 1984), 37, 51–52. Steven J. Zaloga and James W. Loop, *Soviet Tanks and Combat Vehicles, 1946 to Present* (Dorset, England: Arms and Armour Press, 1987), 65–69.

9. Zaloga, 97–103. The BMP had a crew of three and could carry eight infantrymen. Both the BTR and BMP series were amphibious, but that characteristic was not used in this campaign.

10. Siŕen, 118, 122. Thomas, 173. Oliker, 6–9. The Soviets had used overwhelming force to successfully subjugate people in the past without large numbers of casualties. Unrest in Hungary in 1956, Czechoslovakia in 1968, and even the riots in Moscow in the 1980s were quickly ended by a show of force. Perhaps the Russians thought they could repeat the performance.

11. Lieven, 100. During the upcoming battles, Russians would resort to hand-drawn maps or use no maps at all.

12. Oliker, 11–12, 16. Siŕen, 120–121. The military leadership was divided and many officers were skeptical at best. The commander of the North Caucasian Military District, Colonel General Aleksej Mityukin, and the second in command of the land forces, Colonel General Eduard Vorobev, refused to take command of the operation. Neither fully believed in the cause or was confident of success.

13. Oliker, 16–17. Lieven, 109. Adding difficulty to the computations was a considerable number of Chechen men who took up arms when Russian troops moved into their local area but resumed their normal routine when the Russians left. The Russians claimed that thousands of *Muhajeddins* from Afghanistan joined the fray. Utterly fantastic reports tell of female snipers in white tights from the

Baltic states. See also Carlotta Gall and Thomas de Waal, *Chechnya: Calamity in the Caucasus* (New York: New York University Press, 1998), 188, 205–206. This source claims some 1,000 Chechen fighters participated in the fight for Grozny.

14. Thomas, 165. Oliker, 17. Gall, 173–174. Some accounts report there were some 40 to 50 T-62 and T-72 tanks. These numbers may reflect tanks captured after the initial assault or vehicles in the countryside and mountains. The Russians too appeared to inflate the numbers in their estimates, which is common in assessing an elusive enemy. The Chechen tanks played no significant part in the war. They were either knocked out, disabled by falling rubble, or simply abandoned.

15. W. Scott Thompson and Donald D. Frizzell, *The Lessons of Vietnam* (New York: Crane, Russak & Company, 1977), 176. Thomas, 187–188. The AK-47 rifle is rather ineffective at 300 meters, but at 100 meters, it is a very good weapon. It is a simple, idiot-proof rifle that is very effective in close combat.

16. Lieven, 109. Thomas, 171.

17. Oliker, 18–19, 20. Thomas 190–191. Chechen Chief of Staff Aslan Maskhadov, from the basement of the Presidential Palace, led the defense of Grozny. The Internet was also widely used to solicit funds and assistance outside Chechnya. Dudayev and other leaders used mobile television stations to override the Russian broadcasts and deliver messages.

18. Oliker, 15.

19. Ibid., 15–16. The airborne troops were mounted in BMDs, a variant of the BMP, and trucks. Some estimates show that one in five Russian vehicles suffered mechanical breakdowns during the overland trek. Russian helicopters had limited night-vision capabilities and only rudimentary navigation equipment. Most pilots had less than 30 flying hours per year.

20. Gall, 173–174. Lieven, 103. The new reinforcements included army forces; naval infantry (marines) from the Pacific, the Northern, and the Baltic fleets; as well as Ministry of the Interior units. *Spetsnaz* were the Russian Special Forces troops.

21. Thomas, 169–170. Oliker, 13–14. Lieven, 109–111. Meanwhile, the two *Spetsnaz* groups south of the city surrendered to the Chechens after having tried to survive without food for several days. President Dudajev moved his headquarters to Shali, 16 miles south of Grozny at this time as well.

22. Oliker, 24–26. Thomas, 177. Lieven, 111. The Russians emulated their World War II experience by forming assault groups. This proved a disappointment, primarily because the hastily assembled units were unable to work effectively together.

23. Gall, 207. Lieven, 111–112. The Chechens scored a major victory at this time when Major General Viktor Vorobyov was killed by a mortar shell. Sergej Kovalev, the Russian Duma's commissioner for human rights and President Yeltsin's adviser on human rights, estimated the number of civilian dead at about 27,000. The Federal Migration Service put the number of displaced persons at 268,000.

24. Thomas, 173, 176–177. Siren, 126, 132. Gall, 174–175, 177. There were anecdotes where Russian soldiers took to selling their arms to the rebels in exchange for alcohol. Looting was rampant.

25. Oliker, 20–21, 23. Lieven, 114. Thomas, 178.

26. Thomas, 190. Oliker, 26.

27. Oliker, 27. Lieven, 120.

28. Thomas, 180, 182–183. Gall, 213–214, 217, 225. Deputy Minister of the Interior Colonel General Anatolij Kulikov was appointed commander of the combined federal forces in Chechnya.

29. Siren, 130. Oliker, 28.

30. Gall, 242, 249. Oliker, 28. Lieven, 123.

31. Oliker, 29. Gall, 265, 271, 276–277.

32. Thomas, 173.

33. Oliker, 14. Siren, 123–124. No combat units were above 75 percent of their nominal strength. About 70 divisions were less than 50 percent of their nominal strength.

34. Thomas, 173–174, 188. Oliker, 8. Siren, 123–124. The average Russian soldier possessed neither the cultural savvy nor the initiative for urban operations. Most of the Russian conscripts were in the army because they had to be and were anxiously awaiting a return to civilian life. Also, there were an enormous number of combat stress-related casualties during the conflict. See also Lester W. Grau and Timothy Thomas, *Russian Lessons Learned from the Battles for Grozny* (Fort Leavenworth, KS: Foreign Military Studies Office, 2000). This source details the post-combat stress issue in detail.

35. Lester W. Grau, *Russian-Manufactured Armored Vehicle Vulnerability in Combat: The Chechnya Experience* (Fort Leavenworth, KS: Foreign Military Studies Office, 1997). This short work examines the placement of hits and highlights the vulnerabilities of Russian armor used in Chechnya. Some destroyed Russian tanks were hit more than 20 times by RPGs.

Chapter 5

Into the Maelstrom: Fallujah, November 2004

After the fall of Baghdad to the American military in 2003, Fallujah remained one of the most violent areas of the country and the heart of the Sunni Triangle. Violent riots, murders, and bombings became a daily occurrence targeting the occupation forces and the Iraqis collaborating with the interim government or Coalition forces. For months, the local Iraqi police and city leaders proved unable to defuse the situation, but based on their assurances of an improving situation the Americans only ventured into the city occasionally. Meanwhile, the resistance grew stronger taking advantage of a weak government as imams and sheiks incited further violence.[1]

Fallujah dates back to the ancient times of Babylonia as a stop along a primary desert road leading west from Baghdad. Situated on the Euphrates River 43 miles west of Baghdad, Fallujah was part of the modern Anbar Province. A small and unimportant town prior to 1947, increased commerce and the introduction of industry caused its population to gradually swell to about 350,000 by 2003. Fallujah measured about three kilometers square and consisted of over 2,000 city blocks with courtyard walls, tenements, and two-story concrete houses separated by squalid alleyways. Laid out in a grid with a few wide boulevards, the six lanes of Highway 10 ran for two miles straight through the center of the city. South of this highway were decrepit factories while to the north were more spacious homes. As in many cities in Iraq at the time, half-completed homes, heaps of garbage, and wrecks of old cars graced every neighborhood. Ironically, the new highway system sponsored by Saddam Hussein had bypassed Fallujah, and the city's importance and population was on the decline.[2] (See Map 25.)

With over 200 mosques, Fallujah was an important center of Sunni Islam in the region and the population showed a great deal of support for the Ba'athists during the era of Saddam Hussein. Most of the inhabitants practiced extreme Wahhabism and were traditionally hostile to all foreigners, meaning anyone not from Fallujah. The city had a well-earned reputation across Iraq as a very rough town, still firmly entrenched in the tradition of the clan. After the fall of Saddam Hussein and the disintegration of the Iraqi army, there were over 70,000 unemployed men in the streets. With no jobs and an uncertain future, many were highly susceptible to the call for active resistance against the American occupiers. Later estimates showed that over 15,000 Iraqi men did just that.[3]

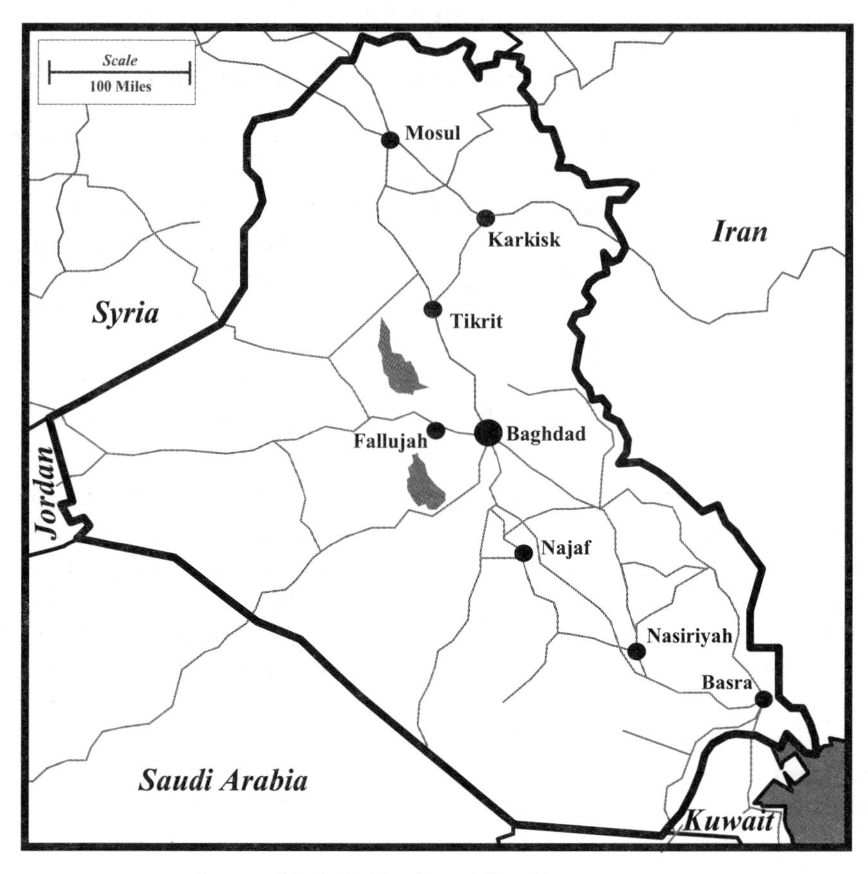

Map 25. The Republic of Iraq.

The American 82d Airborne Division was the first unit assigned responsibility for Fallujah and the surrounding area. Stretched thin over a wide area, the paratroopers were unable to make substantial progress in quelling the unrest. This division was replaced briefly by a 200-man contingent from the 3d Armored Cavalry Regiment in May 2003, but more force was needed. This force was the 2d Brigade of the 3d Infantry Division (Mechanized). Using a carrot and stick approach, there was a noticeable drop in incidents, but Fallujah remained a volatile and dangerous place. Unfortunately, the carrot, in the form of lucrative contracts and lifting of curfews, was often responded to with further attacks by the resistance. The stick was often more effective as the 2d Brigade conducted large-scale sweeps looking for weapons and wanted individuals. The heavy armor of the brigade intimidated the populace, and acts of violence declined a bit more. Meanwhile, efforts to pacify the people of Fallujah by rebuilding infrastructure continued with varying degrees of success.[4]

The resistance fighters in Fallujah were unlike any the American Army had encountered since the Vietnam War. They wore no uniforms and therefore blended in almost perfectly with the population. Operating from their own homes, there was no conventional infrastructure to target such as training camps or bases. Command and control was so loose that there was usually no perceptible chain of command or communications to readily intercept or exploit. Huge stockpiles of weapons and explosive material remained from the war and were readily available for arming new recruits and in the manufacture of improvised roadside bombs. The Wahhabi imams urged members of the resistance to drive out what they saw as the infidel invaders and any Iraqis who collaborated with them. Many mosques became arsenals to stockpile weapons and explosives and safe havens for the resistance. The combination of religious zeal, idleness caused by high unemployment, and hatred for the occupation made recruitment an easy job. The resourcefulness and daring of the fighters made them a deadly foe.[5] (See Map 26.)

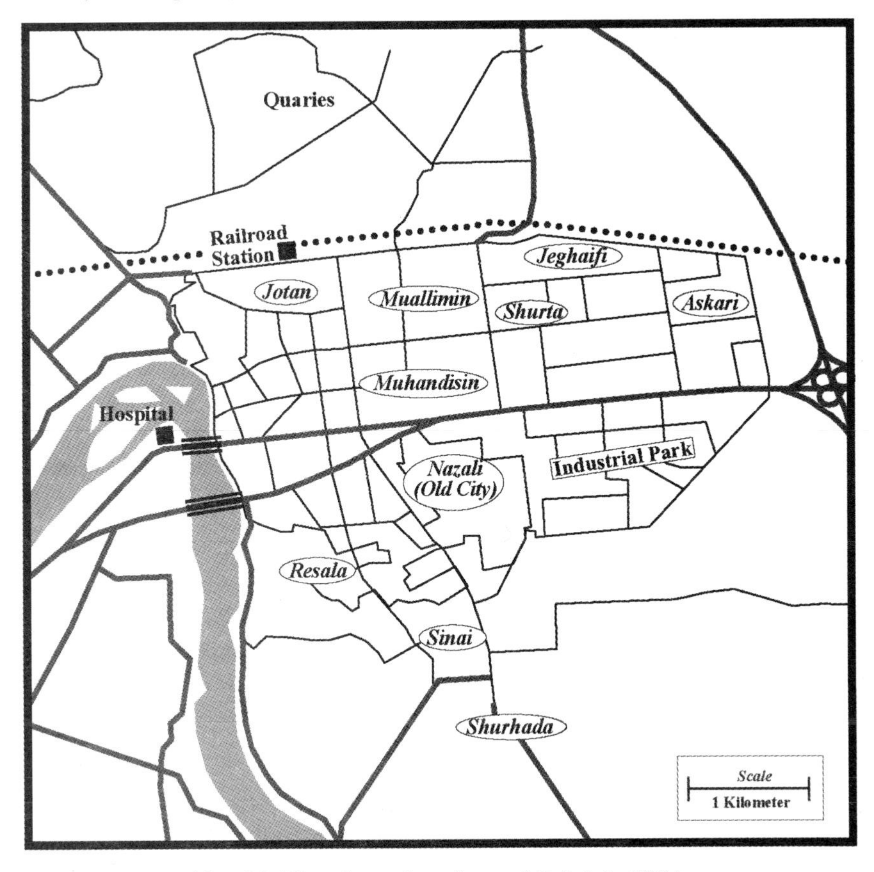

Map 26. The city and sections of Fallujah, 2004.

The 2d Brigade rotated out of Fallujah in August 2003 and was replaced by the 1st Battalion, 505th Parachute Infantry Regiment from the 82d Airborne Division. The situation in the city remained virtually unchanged in spite of the capture of some notable resistance leaders and large numbers of weapons and explosives. Particularly disappointing was the failure of two Iraqi National Guard battalions, arriving in February 2004, to subdue the resistance. Two days after their arrival, a massive attack by the resistance destroyed the central police station as well as the reputation of the guardsmen. The Iraqi battalions were quickly withdrawn in disgrace. There was little significant progress in pacifying the insurgency in Fallujah by the 82d Airborne forces during this rotation. Even the capture of Saddam Hussein on 13 December did not offer a respite; instead, it appeared the resistance grew stronger.[6]

In early March 2004, the 1st Marine Expeditionary Force relieved the 82d Airborne Division in Anbar Province. Instead of focusing on heavy search and sweep operations as the Army units had done, the Marines attempted to shift the focus to emulate their own experience in nation building and winning the hearts and minds of the populace. The Marines hoped the situation would improve by interacting with the people of Fallujah.

The resistance was not impressed. Insurgent leaflets nicknamed the Marines "awat," a sugary soft cake. Attacks escalated. A defining moment came on 31 March 2004 when four contractors were ambushed in Fallujah and their charred corpses were strung up on a nearby bridge. Televised around the world, the scene prompted a heavy response.[7]

In reaction to the murder and mutilation of the four contractors, the Marines and Coalition forces launched Operation VIGILANT RESOLVE on 4 April 2004. The objective of the operation was to pacify and intimidate the violent elements inside Anbar Province, specifically in Fallujah. Four battalions were poised to assault into the city while two more formed a cordon around it. After conducting precision air and artillery strikes, the Marines were prepared to sweep through the city. Senior Marine officers wanted to take a far less drastic approach fearing the heavy damage and Iraqi casualties would be counterproductive to the long-range goal of pacifying the city, however, they were overruled. The Marines began an assault on Fallujah.

On 9 April, after only five days of heavy fighting, the Marines and Coalition forces were ordered to suspend offensive operations in Fallujah to conduct talks with the Governing Council, the city leaders of Fallujah, and representatives of the insurgency. These talks resulted in the delivery of additional supplies to the city by the Iraqi government and the reopening of the Fallujah General Hospital, previously closed because of the siege by the US Marines. The Marines withdrew from the city handing over security responsibilities to the Fallujah Brigade. This light force was composed of former Iraqi soldiers and commanded by Major General Jassim Mohammed Saleh, an officer from the defunct Republican Guards. This cobbled-together unit failed miserably and once again the situation in Fallujah disintegrated. The US Marines maintained a strong ring around the city over the next several months in an effort to contain it.[8]

Over the course of the summer and autumn, the insurgency took the opportunity to recruit personnel and stockpile supplies. Fallujah had become a symbol of resistance and an embarrassment to the Iraqi interim government; at the same time, Coalition forces seemed powerless to do anything about it. Patience was wearing out. City leaders and residents were warned continuously that they were provoking a major assault on the city, but the warning went unheeded. It was generally believed the assault would come soon after the general elections in the United States on 6 November. Those who thought so were right. Beginning in earnest on 30 October, air and artillery attacks pounded select targets in the city as an ominous warning. Near Baghdad, the British Black Watch Regiment relieved American forces preparing for the operation. Power was cut off to Fallujah on 5 November, and leaflets were dropped advising the people who remained in the city to stay inside their homes and not use their cars. On 7 November, the Iraqi government declared a 60-day state of emergency throughout most of the country. Heeding these warnings, between 75 and 90 percent of the civilian population fled the city.[9]

Coalition Forces

The forces surrounding Fallujah and preparing to assault into it were composed of units from the US Army and Marines, supported by aviation assets from the Army, Marines, Navy, and Air Force. Additionally, Iraqi ground forces were to be used in a limited role. In overall command of the operation was Lieutenant General John F. Satter of the US Marines. Satter organized the assault forces into two regimental combat teams, each augmented by two Iraqi army battalions. Total numbers for the operation called for approximately 10,000 Americans and 2,000 Iraqis.[10]

Regimental Combat Team 1 (RCT-1) was assigned to the western half of Fallujah and was composed of three battalions, the 3-1 and 3-5 of the Marines and the 2-7 Armored Cavalry Squadron. Regimental Combat Team 7 (RCT-7) was assigned to the eastern half of the city and was composed of the 1-8 and 1-3 Marines and the 2-2 Mechanized Infantry. In addition to the Army battalions, one Marine tank company augmented each combat team. These M1A2 tanks were widely dispersed down to company level to follow and provide direct support to the Marine riflemen. The 2d Brigade Combat Team (2 BCT) of the 1st Cavalry Division was deployed around the city to block all movement into and out of Fallujah. An Iraqi battalion supported that effort as well.[11]

The officers and soldiers of the units assigned to the operation were composed mainly of veterans of the Iraq War of 2003 and had accumulated a great deal of experience in urban operations. Prior to the operation, these units had the opportunity to train, rehearse, and hone their skills to a sharp edge. The American military forces had up-to-date equipment, including advanced night vision goggles and sights and communications gear. The heavy hitters of the coming operation were the M1A2 Abrams tank and the M2A3 Bradley Infantry Fighting Vehicle. The Marines also employed AAV-7A1 Amphibious Assault Vehicles, but relegated them generally to a heavy weapons platform.

The backbone of the American military's armored forces during this period was the M1A2 Abrams tank. Initially designed in the 1970s, this tank went through a series of refinements and emerged as one of the world's premier armored vehicles. By the time of this operation, the M1A2 had seen combat in the Persian Gulf War in 1991 and was the centerpiece for the Iraq War of 2003 in the drive to Baghdad. In both wars, it far outclassed the Soviet-made T-55 and T-72 tanks. It had a fearsome reputation for its lethality and its ability to take an enormous amount of battle damage and still keep fighting. The M1A2 Abrams weighed over 60 tons, but its gas turbine engine gave it a phenomenal ability to accelerate quickly to its cross-country speed of over 30 miles per hour. Its main armament was a 120mm-smoothbore cannon capable of engaging and destroying targets beyond 3,000 meters. Secondary armaments included a 7.62mm coaxial machine gun and another above the loader's hatch. The commander's cupola was armed with a .50-caliber heavy machine gun. The tank featured a stabilization system to allow the gun to fire while on the move, and the advanced fire control systems were precise enough to engage targets beyond 3,000 meters. A thermal sight allowed for firing at night, through smoke, and during periods of low visibility. Some of the M1A2 Abrams had been further modified with a systems enhancement package, which refined the fire control system and added digital components for communications and a computer map display.[12]

The primary purpose of the M2A3 Bradley, a tracked infantry-fighting vehicle, was to transport infantrymen into battle and then provide supporting and covering fires with an array of on-board weapons. It had a crew of three and could carry six fully equipped infantrymen. A reliable and capable vehicle during the Persian Gulf War in 1991, the M2A3 had received a number of upgrades in armor, fire control, and communications since that time. The main armament of the Bradley IFV was a 25mm chain gun mounted in the turret, capable of firing armor piercing or high explosive rounds at a rate of over 200 rounds per minute. Coaxially mounted to the chain gun was a 7.62mm machine gun, and mounted on the side of the turret was a two-round reloadable tube-launched, optically tracked, wire guided (TOW) missile launcher. The original hull of the Bradley was welded aluminum, but later upgrades included additional steel armor and provisions for reactive armor plates.[13]

The operation also included six small battalions from the growing Iraqi army and security forces. These uniformed forces were armed with AK-47 rifles and Soviet-made machine guns, which were standard in the old Iraqi army. Their training in complex urban operations was rudimentary, so they were to play a supporting role, sweeping through and securing buildings after the Americans had passed through. The Iraqi forces were also ideal to battle insurgents holed up in mosques, because widespread civilian hostility was expected if American forces were to do so. Use of Iraqi forces would also serve to reinforce the perception that Iraqis were able and willing to secure and run their own country.[14]

The Plan of Attack

Originally called Operation PHANTOM FURY by the Americans, the attack on Fallujah was renamed Operation al-Fajr (DAWN) by Iraqi Prime Minister Ayad Allawi. This was a means for the interim Iraqi government to establish control over the city, to bolster its flagging prestige, and to create enough security to hold the national elections scheduled for January 2005 as planned. A secondary but very important objective was to destroy the resistance, killing as many insurgents as possible with a minimum amount of casualties to Coalition forces and civilians. The operation did not receive full support within the Iraqi government and posed a serious risk of alienating large segments of the population, particularly those of the Sunni sect.[15]

The tactical plan was simple. With the cordon in place, the assault forces would assemble to the north of Fallujah and would attack due south within assigned sectors. Boldly breaking with tradition, the US Army's heavy armor would lead the assault into the city with the infantry and Marines closely following to provide cover and to clear each building. Trailing Iraqi forces would conduct further searches for insurgents and logistics caches and assault mosques as needed. The massing of such a force was difficult to accomplish unnoticed, so surprise was achieved by a 12-hour bombardment and activity to the south to draw attention to that sector. The units participating in the operation were to be methodical in their operation, clearing their zone entirely of insurgents. A high level of collateral damage was expected, but civilian casualties were to be avoided if possible. Blocking positions around the city would prevent the escape of hostile combatants. The initial objective was Phase Line Fran, Highway 10, running through the heart of the city. Once the city was secured north of that line, the Coalition forces would fight on to Phase Line Jena to the south. Once that objective was reached, the attacking forces would turn about and sweep through the city in a northerly direction. The attack was scheduled to begin on 7 November.[16]

Over the preceding months, intelligence efforts collected a large amount of information on the insurgent forces in Fallujah. Using every conceivable asset, including Special Forces, human intelligence, unmanned aerial vehicles (UAVs), and satellites, a clear picture of the situation became known. Safe houses, weapons caches, and the routines of key leaders were identified, as well as an approximate number of insurgents active in the city. This information, detailed maps, and overhead imagery were disseminated down to the lowest commander. The commanders and troops at all levels felt confident they knew where the enemy was and could plan their operation in detail. During the operation, these intelligence assets quickly switched to target acquisitions and were instrumental in bringing effective supporting fires on target.

The intelligence picture in November showed the resistance had used the preceding months to turn Fallujah into a fortress. In the city there were approximately 3,000 insurgents, of which about 20 percent were foreign Islamic militants armed with AK-47 rifles, RPG-7s, and a large amount of grenades, mines, and explosives. Attackers could expect a fanatical defense from every building and crevice and from any angle. The dreaded improvised explosives and booby traps were no doubt in place. For movement, the insurgents had dug tunnels between buildings and used the existing sewer system. Believed to be in the city was Jordanian terrorist Abu Musab al-Zarqawi, a high-ranking member of the al-Qaeda terrorist organization. His capture or death was a high priority.[17]

The Assault

Operation DAWN commenced on 7 November as the planned aerial and artillery bombardment began and the ground forces quickly moved to their assault positions. At 1900, the Iraqi 36th Commando Battalion quickly captured the Fallujah General Hospital to the west of the city, while the Marine 3d Light Armored Reconnaissance Battalion secured the two bridges south of the hospital. This mission succeeded in blocking the routes of egress to the west and secured the hospital for use in treating civilian casualties. The major ground assault was now set to begin.[18] (See Map 27.)

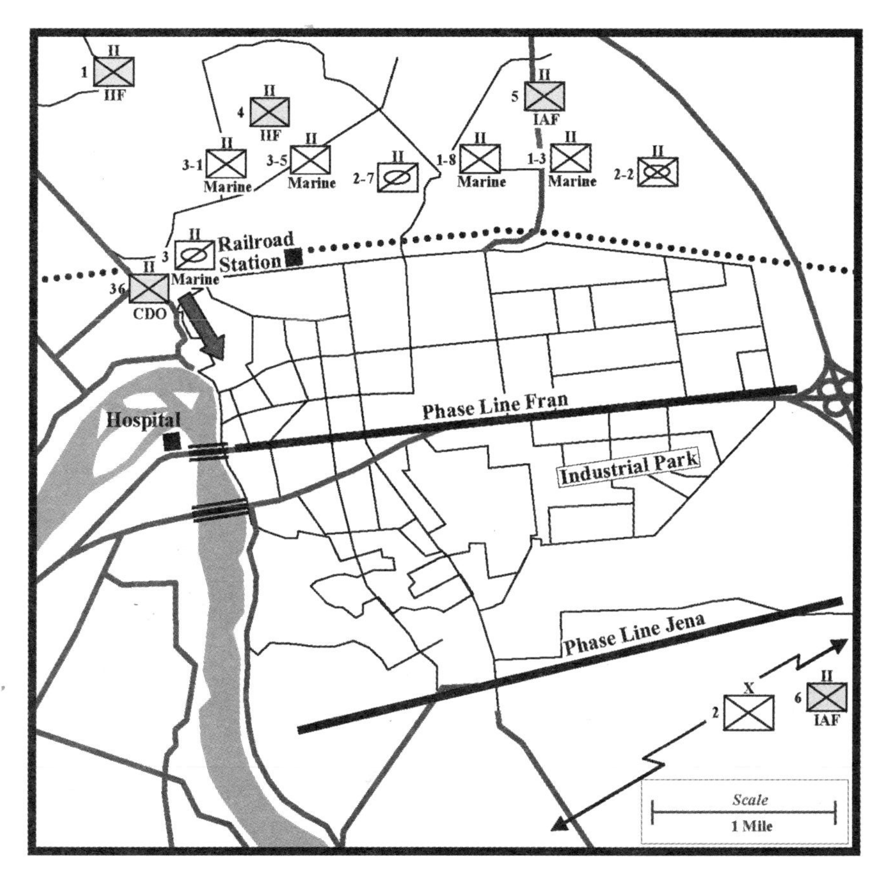

Map 27. In position to assault, 8 November 2004.

The four Marine and two Army battalions began their assault along a broad front in the early hours of 8 November. The heavy armor of the 2-7 Cavalry and the 2-2 Mechanized Infantry led the way into the city. The tanks and IFVs stayed close to either side of the street when possible to provide cover for the vehicles on the other side. Dismounted soldiers provided cover against insurgents attempting to ambush the vehicles by using copious amounts of automatic fire and snipers and by sweeping through the buildings. Often the infantrymen identified strongpoints for the armor, which then used heavy ordnance against the target. Because the tanks' main guns had a limited elevation, armored vehicles were also located in the rear to cover the advance, as their increased distance from targets allowed them to shoot higher than the forward armor. Artillery, mortars, and air strikes eliminated the more stubborn pockets of resistance. Engineers and armored vehicles rammed through the many obstacles and roadblocks. Soldiers and Marines generally entered houses only after tanks blasted through walls or specialists used explosives to create openings. The advance was steady, almost rapid, as the well-trained and equipped Americans ripped through the city. By the afternoon, they had secured the train station and had entered the Dubat and Naziza districts in the west and the Askari and Jolan districts in the east. A seized apartment complex in the northwest looked down on the city, and weapons emplaced there provided excellent fire cover to assaulting forces. The Iraqi forces joined in the attack and aggressively conducted their operations.[19] (See Map 28.)

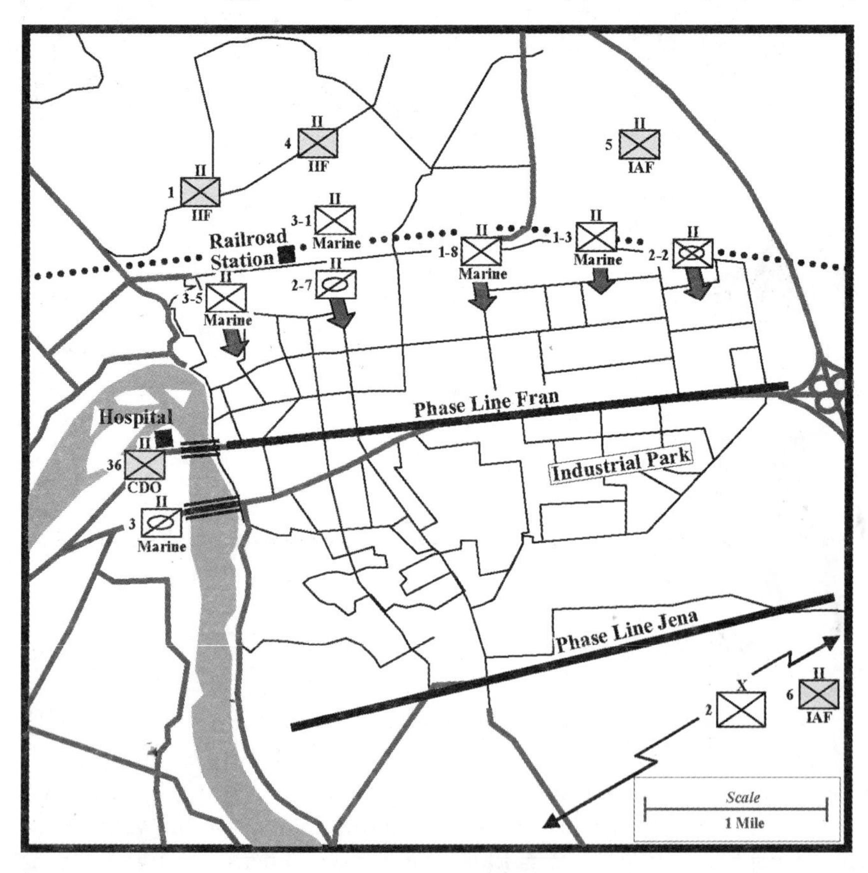

Map 28. Initial assault on Fallujah, 9 November 2004.

The insurgents were clearly overwhelmed from the onset by the speed and shock of the massed armor and firepower. For example, the resistance in the Jolan district in the western edge of Fallujah was expected to be particularly intense. Intelligence reports indicated the most hard line units were located there, and the area consisted of densely packed buildings and narrow streets. Although there was stiff fighting, it was below expectations encountering only small bands of less than 20 insurgents each who were quickly destroyed or forced to fall back under heavy fire. Some analysts believe it was indicative that many insurgents chose to flee the city when they had the chance, or that the deception operation to the south was successful.[20]

In the early morning of 9 November, the Marines conducted a passage of lines through the sectors of the 2-7 Cavalry and 2-2 Mechanized Infantry, placing the armor to the rear of the advance but ready to respond when needed. The Marine tanks remained close behind the advancing riflemen to provide direct fire support. The fighting was so intense, the Army tanks and IFVs could not respond to all the calls for assistance. In those cases, the Marine riflemen had to rely on their organic systems, such as the AT-5 antitank rocket, indirect fires, or air strikes. However, at one point the air strikes and artillery were halted. So many troops were engaged in the densely packed city that a pause was needed to ascertain precise friendly positions to prevent fratricide. By the end of the day, the Army and Marine forces were deep into Fallujah. The greatest gains were in the northeast part of the city, where the 2-2 Mechanized Infantry reached Phase Line Fran, thus cutting the highway, blocking an insurgent escape route, and securing a shorter supply route for Coalition forces.[21]

Heavy fighting continued on 10 November and featured the capture of two large mosques by Iraqi forces. Each of these had been used as insurgent command posts, supply depots, ammunition dumps, and improvised explosive device factories. They had also been insurgent safe houses and fortresses from which to attack Coalition forces. The Iraqi forces found remnants of the black outfits and masks routinely worn by the resistance, as well as banners of the insurgency and videos of the executions of foreign hostages. In addition, many weapons and large amounts of ammunitions and supplies were uncovered. By the end of the day, the Americans could claim that over half of Fallujah was taken, including many key civic and military buildings. Mop-up operations continued in each zone and the Jolan district was turned over to the Iraqis. The fight for the rest of the city lay ahead.[22]

By 11 November, the strategy of attacking and clearing in zone had driven most insurgent forces into the southern part of the city. Coalition forces paused the advance briefly to consolidate and resupply, but the clearing operations continued. By the end of the day, the offensive continued across Phase Line Fran with the armor of the 2-7 Cavalry and 2-2 Infantry again in the lead. The assault was a repeat of the previous days. Full control of Fallujah was expected within 48 hours with an additional week or so to fully clear the city. By 11 November, at least 18 Americans and 5 Iraqi soldiers were killed and about 164 were wounded. An estimated 600 insurgents were killed.[23] (See Map 29.)

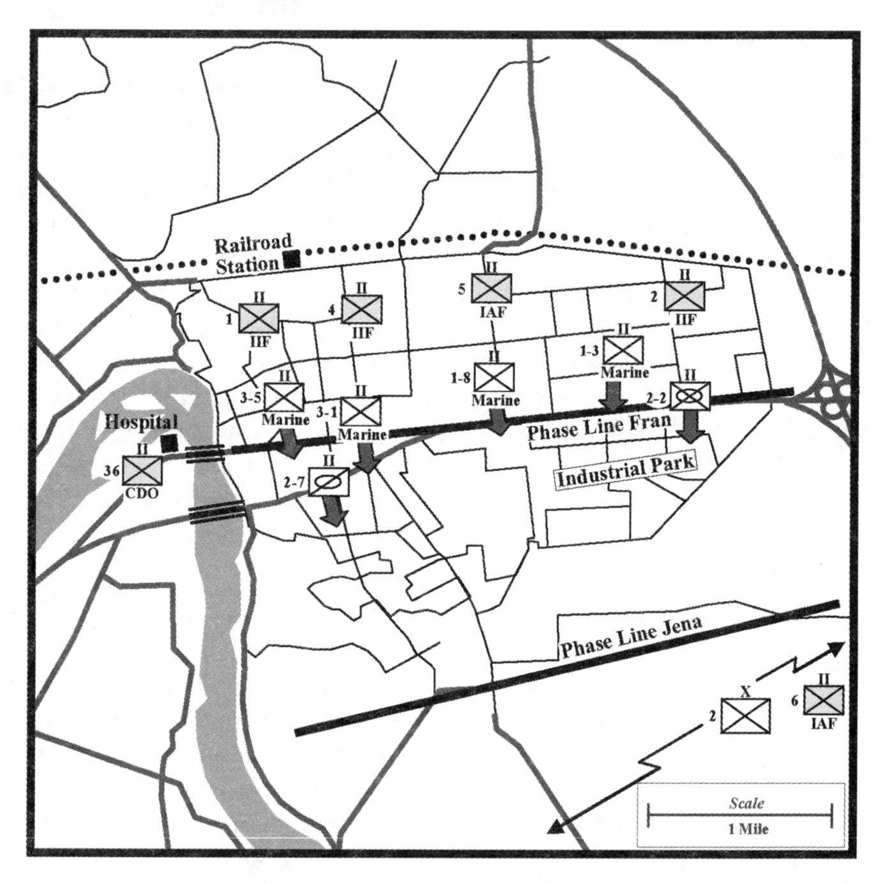

Map 29. Fallujah, 11 November 2004.

The intense street fighting continued for three more days until Coalition forces reached Phase Line Jena in the south. Over 300 insurgents surrendered, many having been surrounded in a mosque. Thousands of AK-47s, RPGs, mortar rounds, and improvised explosive devices were found in houses and mosques. There were still fears though that sleeper cells would rise up once the assault had passed through an area.[24]

When Coalition forces reached Phase Line Jena on 15 November, they turned about and began re-clearing buildings as they moved northward. The Army and Marine battalions broke down into company, platoon, and squad-size elements to thoroughly search for hiding insurgents and caches. The progress was methodical with a great concern for booby traps laid by roving bands of the resistance. The efforts were not in vain as additional weapons and explosives were found. By 16 November, the city of Fallujah was declared secured by Coalition forces, although the search and sweep operations continued for several weeks. (See Map 30.)

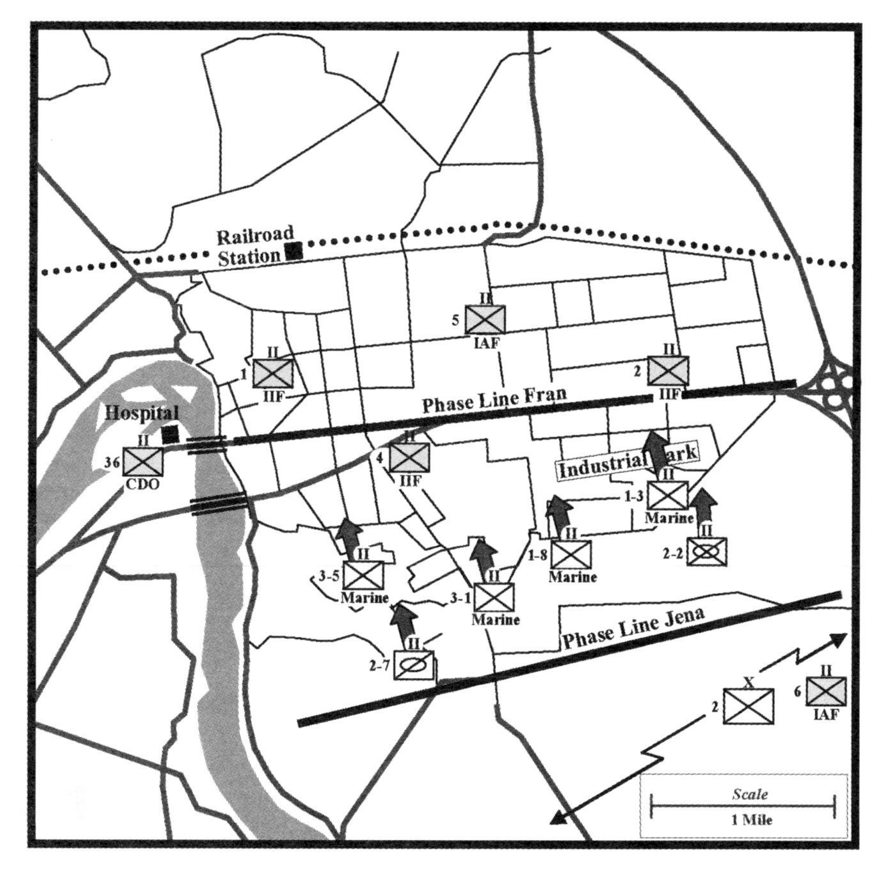

Map 30. Fallujah, 15 November 2004.

Operation DAWN resulted in the death of 38 US troops, 6 Iraqi soldiers, and between 1,200 and 2,000 insurgents. Three of the American fatalities were nonbattle related injuries. At least 275 Americans were wounded. Between 1,000 and 1,500 insurgents were captured.[25]

The Dust Settles

The operation heavily gutted much of the city. Many reports indicate that over 60 percent of Fallujah's buildings were damaged, 20 were destroyed outright, and 60 of the mosques were heavily damaged. In response to the operation and damage, the Sunni minority in Iraq was enraged. Insurgent activity surged across the country and demonstrations were widespread. Sunni turnout was indeed low in the January elections, but they were held. However, subsequent elections in June and December 2005 saw increased Sunni participation.[26]

The Iraqi government sent medical and reconstruction teams to the area with 14 trucks loaded with medical supplies and humanitarian goods. Unable to enter the city because of the military operations, they were diverted to villages surrounding Fallujah where tens of thousands of displaced civilians had fled to escape the conflict. Meanwhile, Iraqi and American forces sought out civilians in need of medical care using loudspeakers, leaflets, and word of mouth. The Fallujah General Hospital was available and ready for use.[27]

Residents of Fallujah were allowed to return in mid-December and the slow process of reconstruction began. It remained an enclave of the resistance, but its strength was greatly weakened and Operation DAWN served as an example for cities in open defiance of the Iraqi government.[28]

In Retrospect

By November 2004, the American military were highly proficiency in the tactics and techniques of urban warfare. Many if not most of the officers and troops were veterans of the Iraq War of 2003 and the subsequent occupation of the country. The US Army and Marines had standing urban operations doctrines, which they applied and modified to meet the situation. Soldiers and Marines had individual styles. The Army troops were inclined to be more methodical in tactics but liberal in the use of heavy ordnance, and the Marines, by tradition, tended to rely on the shock and audacity of their small unit attacks and called on heavy support only after an attack stalled. However, both services overcame organizational friction and worked well together toward a common goal.

Heavy ordnance delivered from aircraft and artillery was used effectively as a rolling barrage to cover the movement of ground forces and to obliterate insurgent strongpoints. Key buildings and mosques were spared when possible, but they were engaged aggressively when insurgents used these structures in their operations. Precision munitions and excellent communications ensured quick, accurate, and deadly fires.

Considering the complexities of the situation, the intelligence support for Operation DAWN was superb. Using the prior weeks and months to great effect, the various intelligence agencies and platforms were able to paint an accurate picture of the situation and disseminate that information down to the lowest echelons. When the battle began, these assets quickly shifted to acquiring targets and assessing the capabilities and intentions of the insurgents.

Iraqi forces proved capable of cooperating within the coalition for this operation. Although they played a limited role, they did attack key targets, like mosques, thus avoiding widespread consternation had an American unit done so. The light Iraqi units fought effectively within their capabilities.[29]

If the insurgents were hoping for a replay of the Russian debacle in Grozny in 1994, they were disappointed. The strategy of "defenseless defense" used so effectively there did not work in Fallujah. The American and Iraqi forces were successful in countering this tactic by not rushing to the center of the city to be surrounded and eliminated piecemeal. Instead, they cleared and secured each building and the routes of ingress before moving on to the next. Additionally, some American and Iraqi forces remained behind the advance to keep the insurgents from reoccupying previously cleared areas. Establishing clear zones of operation and excellent communications facilitated this.

A key element in the success of the coalition in Fallujah was the application of American armor, namely the M1A2 Abrams tank. The Abrams was able to take enormous punishment and continue operating. In many instances, these tanks received multiple hits from RPG-7s, which failed to penetrate the heavy armor; even large improvised explosives failed to knock tanks out. Although the actual number is not currently released to the public, contemporary media reports show only two Abrams tanks were destroyed during this bitter battle. The tactics used by the Americans offset the inherent design weaknesses of tanks in the cities. Operating in pairs, tanks covered each other while others remained a short distance behind lending support. The same can be said about the Bradley vehicles, although their armor was far less capable. The Marines had dispersed their tanks to provide direct support to the riflemen, and this time-honored tactic worked to destroy systematically tough enemy positions. Conversely, the Army battalions assigned to this operation used a different approach. Instead, they led their assault with the heavy armor, which blasted through the city and unhinged the enemy defenses. This allowed for the rapid advance of the infantry and the clearing of their zone and ensured a swift victory.[30]

The battle for Fallujah was a stunning victory with a historically low casualty rate for an urban fight of this size. It reaffirmed the capabilities of heavy armor in cities.

Notes

1. A stream of information available to the public was just beginning at the time of this writing. The text and analysis were compiled using various unit briefings, media reports, and the few published works available. No classified material was used in the writing of this work.

2. Mike Tucker, *Among Warriors in Iraq: True Grit, Special Ops, and Raiding in Mosul and Fallujah* (Guilford, CN: The Lyon's Press, 2005), 89–90. Bing West, *No True Glory: A Frontline Account of the Battle for Fallujah* (New York: Bantam Books, 2005), 13. Anbar Province, the largest province in Iraq, borders Syria and Jordan to the west and Saudi Arabia to the south. Anbar is largely a combination of desert and steppe and is generally an inhospitable environment. Temperatures can get as high as 115 degrees Fahrenheit in the summer.

3. West, 13–14. Fallujah was known as the "city of mosques" due to the large number of them.

4. West, 14–16.

5. Anthony H. Cordesman and Patrick Baetjer, *The Lessons of Modern War, Volume I: The Arab-Israeli Conflicts, 1973–1989* (Boulder, CO: Westview Press, 1990), 30–31, 40, 46. The common term for these bombs was improvised explosive devices (IEDs).

6. West 26, 47–49. Cordesman, 72.

7. West, 3–4, 51, 58.

8. Cordesman, 52, 72, 356–357. See also West, 258. This operation was controversial from concept to execution. The US media portrayed it as a humiliating loss and some Iraqi battalions refused to fight. The Fallujah Brigade, a hastily raised and equipped force, was disbanded within four months. Many of its soldiers joined the insurgency. Across the country, the number of significant attacks did decline for a month or so but rose again.

9. Fallujah residents were without running water and were worried about food shortages. Faced by a hostile press and strong Democratic Party opposition, it is no surprise the Bush Administration waited until after the general elections to launch an attack. The timing also avoided the hot summer temperatures.

10. West, 258.

11. West, 258–260. The tactic of establishing a tight cordon around Fallujah had not been employed back in April. For brevity and clarity, the Army and Marine battalions are known by their official designations. They were, in fact, task forces with infantry or armor attached for the mission at hand.

12. R.P. Hunnicutt, *Abrams: A History of the American Main Battle Tank, Volume 2* (Novato, CA: Presidio Press, 1990), 210, 224–225, 229, 274. Only 18 Abrams tanks were lost to enemy action during the Persian Gulf War of 1991, about half by mines. Not a single Abrams crewmember was lost in that conflict. Only a few M1A2s were lost in 1993, and mines again were their deadliest enemy. After 1992 many analysts declared the Abrams tank a legacy system in line for retirement or replacement. Their assessment proved premature.

13. Steven J. Zaloga, *The M2 Bradley Infantry Fighting Vehicle* (London:

Osprey Publishing, 1986), 22–24, 33. William B. Haworth, Jr., "The Bradley and How it Got That Way: Mechanized Infantry Organization and Equipment in the US Army" (Ph.D. diss., Ann Arbor, MI: UMI Dissertation Services, 1998), 205–207, 213. Of over 2,000 Bradley IFVs participating in the Iraq War, only three were lost to enemy action. The TOW missile required the Bradley to stop in order to fire this weapon. The TOW had an effective range of over 3,000 meters.

14. It was expected that over 2,000 Iraqi troops would participate in the operation, but an unknown number of them deserted prior to D-Day.

15. Cordesman, 51, 85, 97. The Industry Minister, Hajim Al-Hassani of the mainly Sunni Iraqi Islamic Party, quit the government in protest over this operation.

16. West, 258. Under international law, mosques are granted protected status but loose that status if used for military purposes. The attack in April had come from the south and the Americans had hinted to the media and Iraqis that this time it would come from that direction as well.

17. West, 257. Improvised explosive devices were the most-feared threat to the coalition planners.

18. Cordesman, 104, 359–360. West, 260.

19. West, 263, 268–269.

20. West, 270.

21. West, 284–285, 315. It is believed that Abu Musab al-Zarqawi, leader of the insurgent faction in Fallujah, fled the city on this day. Meanwhile, the leading Sunni political party, the Iraqi Islamic Party, announced it was withdrawing from the interim government and called for a boycott of the upcoming national elections.

22. Cordesman, 104, 359–360. West, 275. The Iraqi 5th Battalion, 3d Brigade seized Al Tawfiq Mosque and the Iraqi Police Service's Emergency Response Unit and elements of the 1st Brigade of the Iraqi Intervention Force captured the Hydra Mosque. A number of insurgents were captured and transferred to the Abu Ghraib Prison for further questioning. Some armed women and children took part in the fighting throughout the city.

23. West, 282.

24. Cordesman, 360. On 13 November Prime Minister Alawi declared Fallujah liberated. West, 293, 305. During the fight a Marine lieutenant was accused of shooting a wounded insurgent he believed to be feigning death. A media circus ensued, but the young officer was later acquitted of all charges. The court-martial ruling stated he was acting in self-defense.

25. West, 316. Cordesman, 360. Tucker, 94.

26. West, 315–317. These numbers indicate that over a quarter of the buildings and mosques had heavy damage.

27. David L. Phillips, *Losing Iraq: Inside the Postwar Reconstruction Fiasco* (New York: Westview Press, 2005), 216.

28. Cordesman, 103. Phillips, 222. The loss of Fallujah deprived the Sunni insurgents and terrorist groups of their major sanctuary in Iraq.

29. Cordesman, 103.

30. Jason Conroy and Ron Martz, *Heavy Metal: A Tank Company's Battle to Baghdad* (Dulles, VA: Potomac Books, Inc., 2005), 169, 267–268. The US Army's M1127 Stryker vehicle was not used in this fight. It is the author's opinion that the Stryker's armor would have proven too thin, the wheels too vulnerable to enemy fire and debris, and its maneuverability too limited to be effective in such a scenario. A number of reports from Iraq praised this vehicle and its capabilities, but I remain skeptical. There were Stryker units able to participate, but did not. Instead, they were used to maintain the cordon around the city, a role for which they were better suited.

Chapter 6

Conclusion

Urban warfare is a deadly business and a growing prospect for future conflicts as global urbanization trends continue. Furthermore, fighting in the streets is a lucrative tactic for nations and factions who are incapable of fighting more conventionally against a large opponent armed with advanced weaponry.

The previous case studies outline a few of the historical examples of armored forces in urban warfare. A common trait is that all were eventually successful, although to varying degrees. Even the Russian debacle in Grozny was successful at the tactical level, although the entire operation failed in its objective of subduing the resistance. Any battlefield shortcomings found within these historical examples came not from the armor forces but from the planners and leaders at the strategic level. In each case it was the armor forces' firepower that allowed the accompanying infantrymen to close with their opponent and win the day. If one were to remove the armor from these scenarios, the outcomes would have been far more costly in casualties and time.

The doctrine of modern armies had long emphasized the avoidance of tank employment in large urban areas, and after World War II the emphasis of the tank as an infantry support weapon shifted. Instead, the major nations of the world built tanks designed to operate on the battlefield primarily against enemy armor. This is shown in the placement of its armor toward the front of the vehicle, the long-range optics, and a high-velocity main gun designed to defeat opposing tanks. Tanks were also to avoid costly engagements using fire and maneuver techniques against weak points in the enemy defenses. Doubters of this assessment need look no further than the ammunition racks. During the years 1945 to 2003, the vast majority of rounds routinely carried by American tanks were of the hypervelocity sabot or high explosive antitank (HEAT) variety. The sabot round has limited use in the cities. The HEAT rounds are useful against walls and masonry, but do not have the area blast effect of high explosive (HE) rounds or the devastation of a flechette (beehive) round against personnel.

Generally, tanks have a number of characteristics that limit their use in confined areas. Their weaponry is not situated for a close-in fight, particularly to the sides and rear where they cannot engage targets. The main guns are often too long to traverse fully in narrow streets, and the small vision blocks severely restrict target acquisition. Armor is often very thin on the top, flanks, and rear, and extremely thin on the undersides. The latter makes tanks exceptionally vulnerable to mines and improvised explosive devices so common in a city fight. Therefore, the very design that makes a tank a feared weapon on the open battlefield renders it extremely vulnerable in the close confines of a city. It is no wonder leaders and planners want to avoid committing large armor forces to urban battles. Nevertheless, armies have employed armor in cities over the years because despite their potential shortcomings the tank is the most effective all-weather system to bring precision heavy ordnance to the target.

The battle of Aachen was a fine example of using armor to support the infantry in the urban fight. By 1944, the Americans were veteran troops, and although not specifically trained for urban operations, these forces had developed the necessary tactics and skills to attack the fortifications of the German *Westwall.* Armor forces were also adept at cooperating with supporting infantry and fire support at that stage of the war. These combined experiences were quickly and successfully adapted to the city fight. The command and control displayed by the Americans was methodical, but effective in coordinating the battle. The vulnerable M4 Sherman tanks and M10 tank destroyers were effective as long as their infantry support shielded them from the dreaded *panzerfausts*. The isolation of the German defenders and the rapid movements of Task Force Hogan disrupted the German defense and speeded the conclusion of the fight for Aachen.

The employment of armor in the fight for Hue was a result of the fortunate proximity of Marine M48 tanks at the onset of the Tet Offensive. Able to withstand enemy B-40 rockets, the heavy armor was able to bring its firepower to bear in support of the Marines' efforts to retake the city. Untrained for an urban fight, the Marines adapted quickly to overcome poor weather, political constraints, and a determined enemy. They also worked closely with the ARVN forces, which were badly shaken in the initial assault and slower to recover. The US and ARVN armor proved decisive in supporting the riflemen's efforts to retake Hue. Unable to destroy the heavy armor with the B-40s, the Marines and ARVN troops closed in for the kill and subjected the *NVA* forces to heavy fire.

The campaign for Beirut saw a slight shift in the tradition of armor in the supporting role of the infantry in urban terrain. Possessing an order of battle designed to fight in the open desert, the Israelis proved adept at reorganizing their forces and modifying their command and control procedures in this operation. Assigning infantry officers to command the armor-heavy units was one dramatic example. In the drive up the coast and into Beirut, it was often the tanks leading the assault with the infantry closely following in support. The Israeli tanks blasted through the *PLO* strongpoints and refugee camps and quickly reduced the cities of Tyre and Sidon using massed firepower and speed. Bold daring and the extensive use of engineer assets to build bridges and to clear mines and debris maintained this momentum. The capable Merkava and M60 tanks were far superior to their opposite number and were generally able to withstand multiple hits from the venerable RPG-7. Political concerns beyond the tactical commanders' control blunted this armored drive.

The initial assault into Grozny is probably the definitive example of a poorly executed armored assault into a large urban area. Untrained and poorly led, the Russian armor crews were easy prey to the Chechen RPG-7 gunners. Dozens of Russian tanks and APCs were quickly destroyed in the narrow streets after they were immobilized and unable to fight back due to the technical limitations of the vehicle design. The massive application of heavy firepower ultimately achieved success, but reduced the city of Grozny to rubble in the process.

The November 2004 battle for Fallujah is perhaps the most successful use of heavy armor in an urban environment. Well-led veteran forces were able to crush quickly their opposition by boldly using tanks and APCs to spearhead the assault. The shock and speed proved great enough to disrupt the enemy defense, which was never able to recover to make a coordinated stand.

The preceding chapters show the success of the tank in urban warfare; even the Russians in Grozny were ultimately successful once they adapted to the unexpected situation. The units participating in the initial operations did not practice the established Russian urban warfare doctrine in 1994. The malaise of the post-Cold War was acute in the Russian Army, and units were not prepared for what awaited them. The results were disastrous. The Americans in 1944 and 1968 did not have an established doctrine for fighting in large cities, but the use of veteran troops in the operations negated this deficiency. Learned in the hedgerows and jungles in previous fights were the small unit teamwork and tactical skills needed for the city fight. With a bit of ingenuity and drive, these attacking units prevailed.

Ideally, the preparation for urban combat begins during peacetime. Various scenarios, options, constraints and limitations, legal factors, and city characteristics must be studied and understood. Leaders and planners must remember that each urban operation will be unique and there is simply no standard urban operation as no two cities are alike. There is just too much variance with physical layouts, enemy forces, and civilian demographics. Doctrine and unit training require addressing specific skills to fight in the cities, but must remain flexible enough to adapt quickly to a host of possible situations. Considering historical trends, the fight for the streets is often decided at the crew and platoon level. Yet there exist political considerations, legal limitations, infrastructure, and evolving enemy methods that are beyond the ability of the junior leaders to research and incorporate into their training and operations. Understanding the nature of urban warfare is a difficult task for any army, and the tasks required to sufficiently sustain or support urban combat are enormous. With one possible exception, all of the examples in this book show armies attacking cities without extensive training in urban operations. Each was ultimately successful, either because of a high degree of skill or experience in small unit tactics or a large application of heavy firepower. To bridge the gap between peacetime training and commitment to the battle on the street, the US Army must fully embrace the concept of using armor in urban warfare and prepare accordingly.

This historical narrative shows that the mobility of tanks and armored vehicles, along with their protective armor, allowed the delivery of their heavy firepower into the fight for the cities. Maintaining mobility is vital to the effectiveness of tanks and to their survival. The obvious tactic is to provide a robust engineer effort to clear debris, rubble, and armor hulks, and to eliminate mines and improvised explosives from the paths of advancing armor. The most graphic example of the failure to do so was the Russians in Grozny. Once trapped in the narrow streets, even the advanced T-80 tanks were vulnerable to the simple RPG-7. If stationary, even the most capable tank becomes a pillbox with limited angles of fire in narrow streets and alleys. The fight for Fallujah is a dramatic example of a rapid armored advance that unhinged the enemy's defense and allowed a rapid victory.

An effective means to maintain mobility is to disrupt the enemy's fire plan by the application of maneuver, firepower, or obscurants. A high tempo operation would also challenge an enemy's ability to react and engage. These methods require an intimate knowledge of the capabilities of units and weapons on urban terrain. Commanders and staffs must understand the advantages and disadvantages urbanization offers and its effects on tactical operations. An operation based on maneuver could avoid one based on an attrition strategy and prevent heavy friendly losses.

Conducting a high tempo operation over a sustained period will require a massive logistical effort, particularly in the resupply of ammunition. Fuel is also of great concern, as the consumption rate of the gas turbine engines found in many modern tanks is notoriously high. In the battle for Hue, the Marines had no such plan and were forced to withdraw their tanks from the front lines to refuel and rearm. The riflemen fighting their way to the Citadel keenly felt the temporary loss of this combat power. Flexible command and control and an effective intelligence operation are also vital elements in a highly fluid battle for the streets.

Vehicle design for tanks and armored vehicles have traditionally focused on the conventional fight in open terrain. Armor protection on the top, rear, and underside is very thin in comparison to the frontal arc, and vulnerable to the short-range rocket grenades and mines so common in urban battles. Armor plate must be able to absorb or deflect the damage caused by the enemy's weapons. If unable to do so, armored vehicles become death traps for crewmembers and their heavy weapons neutralized. The RPG-7 rocket propelled grenade is legion throughout the world and US tank crews will encounter it for years to come. Countries producing this weapon have taken note of the effectiveness of laminate armor and are undoubtedly searching for a more potent warhead for the RPG-7 or an even more advanced hand-held weapon. The recent tactic of boldly leading urban assaults with tanks, as in Fallujah, may come to an abrupt end and require returning armor to the infantry support roles as was seen in Aachen and Hue.

Since a new tank or armored vehicle designed specifically for the city fight will not arrive any time soon, the US Army must rely on its current inventory. Whether a new vehicle arrives on the scene or a legacy system is used, there is one unshakable principle in their employment in urban terrain. Except in the most extraordinary circumstances, tanks and armored vehicles must be closely supported by sufficient infantry or massed firepower to protect them from a wide variety of hand-held antitank weapons common on the modern battlefield. Vehicles cut off from their infantry support will quickly fall victim to their enemy. The battle for Aachen is perhaps the best example of this. The M4 Sherman was a good infantry support vehicle, but the German antitank weapons were devastating against its inadequate armor. To compensate, the Americans made great efforts to screen them with infantry. The Russian operation in Grozny was the other extreme of this discussion point. The initial sortie into the heart of the city was decimated as the accompanying infantry remained mounted and unable to shield the tanks from the rain of RPG-7 fire. In Operation Peace for Galilee, the Israelis realized the organization of their forces was not suitable for the fight and rearranged their units and command structure beforehand.

Military operations in large cities will continue to be more common as the world urbanizes and more deadly with the introduction of new weapons.

The US Army's basic urban warfare doctrine is sound; nevertheless, it is being refined by the experiences learned in Iraq. Currently these lessons are ingrained through the ranks as personnel serve on multiple deployments and as training centers and service schools incorporate these lessons into their curriculums. For the short term, the United States possesses the premier fighting force for the task of taking large urban centers.

At the time of this writing, there is a general worldwide stagnation in new tank design and fielding. The major powers of the world are focused on extending the service life of their existing fleets with various upgrades in firepower, mobility, and armor protection. For the United States, the next generation tank is very much a work in progress as new technologies are developed. Until that time, US planners will need to "make do" with the Abrams series tanks, as there is nothing more effective to bring precision heavy fire into the cities. It is unlikely that a single technology or system will emerge in the near future that will swing the balance to the attacker in the cities. Instead, an effective solution to the urban fight will only be attained through the integration of strategic concepts, doctrine, operational needs, technological advances, system design, and the appropriate organization of command, control, training, and education. No doubt tanks and armored vehicles will play a vital role.

About the Author

Kendall D. Gott retired from the US Army in 2000 after serving as an armor, cavalry, and military intelligence officer. His combat experience consists of the Persian Gulf War and two subsequent bombing campaigns against Iraq. A native of Peoria, Illinois, Mr. Gott received a B.A. in history from Western Illinois University in 1983 and a Masters of Military Art and Science degree from the US Army Command and General Staff College. Before returning to Kansas in 2002, he was an adjunct professor of history at Augusta State University and the Georgia Military College. In October 2002 he joined the Combat Studies Institute where he researches and writes articles and studies on topics of military history. His book-length works include *In Glory's Shadow: The 2d Armored Cavalry Regiment During the Persian Gulf War, 1990-1991*, and *Where the South Lost the War: An Analysis of the Fort Henry–Fort Donelson Campaign, February 1862*. Mr. Gott is a frequent speaker at Civil War roundtables and appeared on a recent History Channel documentary on the Battle of Mine Creek, Kansas, and the documentary *Three Forts in Tennessee* by Aperture Films.

Bibliography

Government and Doctrinal Publications

Colby, Elbridge. *The First Army in Europe*. US Senate Document 91–25. Washington, DC: US Government Printing Office, 1969.

Department of the Army. FM 3-06, *Urban Operations*. Washington, DC: US Government Printing Office, June 2003.

Department of the Army. FM 3-06.11, *Combined Arms Operations in Urban Terrain*. Washington, DC: US Government Printing Office, 28 February 2002.

Department of the Army. FM 90-10, *Military Operations in Urban Terrain (MOUT)*. Washington, DC: US Government Printing Office, 1979.

Department of the Army. FM 90-10-1, *An Infantryman's Guide to Combat in Built-up Areas*. Washington, DC: US Government Printing Office, 1993.

General Service Schools. *The Employment of Tanks in Combat*. Fort Leavenworth, KS: General Service Schools Press, 1925.

Grau, Lester W. and Timothy Thomas. *Russian Lessons Learned from the Battles for Grozny*. Fort Leavenworth, KS: Foreign Military Studies Office, 2000.

Grau, Lester W. *Russian-Manufactured Armored Vehicle Vulnerability in Combat: The Chechnya Experience*. Fort Leavenworth, KS: Foreign Military Studies Office, 1997.

Killblane, Richard E. *Circle the Wagons: The History of US Army Convoy Security*. Fort Leavenworth, KS: Combat Studies Institute Press, 2005.

MacDonald, Charles B. *US Army in WWI: The Siegfried Line Campaign*. Washington, DC: US Government Printing Office, 1984.

Department of Defense. *Handbook for Joint Operations*. Washington, DC: US Government Printing Office, 17 May 2000.

United States War Department. *Infantry Field Manual*. Washington, DC: US Government Printing Office, 1931.

Books and Secondary Sources

Anderson, Jon L. *The Fall of Baghdad*. New York: The Penguin Press, 2004.

Bodansky, Yossef. *The Secret History of the Iraq War*. New York: Harper Collins Publishers, 2004.

Butler, William and William Strode, eds. *Chariots of Iron: 50 Years of American Armor.* Louisville, KY: Near Fine Harmony House, 1990.

Conroy, Jason and Ron Martz. *Heavy Metal: A Tank Company's Battle to Baghdad*. Dulles, VA: Potomac Books, Inc., 2005.

Cordesman, Anthony H. *Iraqi Security Forces: A Strategy for Success*. Westport, CN: Praeger Specialty International, 2006.

Cordesman, Anthony H. and Patrick Baetjer. *The Lessons of Modern War, Volume I: The Arab-Israeli Conflicts, 1973–1989*. Boulder, CO: Westview Press, 1990.

Cooper, Belton Y. *Death Traps: The Survival of an American Armored Division in World War II.* Novato, CA: Presidio Press, 1998.

Dewar, Michael. *War in the Streets: The Story of Urban Combat from Calais to Khafji.* Newton Abbot, UK: David & Charles, 1992.

Dupuy, Trevor N. *The Evolution of Weapons and Warfare.* Fairfax, VA: Hero Books, 1984.

Eshel, David. *Chariots of the Desert: The Story of the Israeli Armoured Corps.* London: Brassey's Defense Publishers, 1989.

________. *Mid-East Wars: Israel's Armor in Action.* Hod Hashron, Israel: Eshel-Dramit, 1978.

Evangelista, Matthew. *The Chechen Wars: Will Russia Go the Way of the Soviet Union?* Washington, DC: Brookings Institute Press, 2002.

Fowkes, Ben, ed. *Russian and Chechnya: The Permanent Crisis.* New York: St. Martin's Press, Inc., 1998.

Fuller, John F.C. *Armored Warfare.* Westport, CN: Greenwood Press, 1994.

Gall, Carlotta and Thomas de Waal. *Chechnya: Calamity in the Caucasus.* New York: New York University Press, 1998.

Guderian, Heinz G. *From Normandy to the Ruhr: With the 116th Panzer Division in World War II.* Bedford, PA: The Aberjona Press, 2001.

Hammel, Eric M. *Fire in the Streets: The Battle for Hue, Tet 1968.* Chicago, IL: Contemporary Books, 1990.

Haworth, William B., Jr. "The Bradley and How it Got That Way: Mechanized Infantry Organization and Equipment in the US Army." Ph.D. diss, Ann Arbor, MI: UMI Dissertation Services, 1998.

Herzog, Chaim. *The Arab-Israeli Wars: War and Peace in the Middle East.* New York: Vintage Books, 1984.

Himes, Rolf. *Main Battle Tanks: Development in Design Since 1945.* Washington, DC: Brassey's Defense Publishers, 1987.

Hunnicutt, R.P. *Abrams: A History of the American Main Battle Tank, Volume 2.* Novato, CA: Presidio Press, 1990.

________. *Patton: A History of the American Main Battle Tank.* Novato, CA: Presidio Press, 1984.

________. *Sherman: History of the American Medium Battle Tank, Volume 2.* Novato, CA: Presidio Press, 1978.

Hogg, Ian V. *Armour in Conflict: The Design and Tactics of Armoured Fighting Vehicles.* London: Jane's Publishing Company, 1980.

________. *Tank Killing: Anti-Tank Warfare by Men and Machines.* New York: SARPEDON, 1996.

Jarymowycz, Roman J. *Tank Tactics: From Normandy to Lorraine.* London: Lynne Rienner Publishers, 2001.

Kaplan, Philip. *Chariots of Fire: Tanks and Their Crews.* London: Aurum Press Ltd., 2003.

Karnow, Stanley. *Vietnam: A History.* New York: Penguin Books, 1983.

Knickerbocker, H.R. *Danger Forward: The Story of the First Division in World War II.* Washington, DC: Society of the First Division, 1947.

Lieven, Anatol. *Chechnya: Tombstone of Russian Power*. New Haven, CN: Yale Univeristy Press, 1998.

Morrison, Wilbur H. *The Elephant and the Tiger: The Full Story of the Vietnam War.* New York: Hippocrene Books, 1990.

Nolan, Keith W. *Battle for Hue: Tet, 1968*. Novato, CA: Presidio Press, 1983.

Oberdorfer, Don. *Tet!* Garden City, NY: Doubleday & Company, Inc., 1971.

Ogorkiewicz, Richard M. *Design and Development of Fighting Vehicles*. Garden City, NY: Doubleday, 1968.

________. *Technology of Tanks*. Surrey, England: Jane's Information Group, 1991.

Oliker, Olga. *Russia's Chechen Wars, 1994–2000: Lessons from Urban Combat*. Santa Monica, CA: RAND Corporation, 2001.

Perrett, Bryan. *Iron Fist: Classic Armoured Warfare Case Studies*. London: Arms and Armour Press, 1995.

Phillips, David L. *Losing Iraq: Inside the Postwar Reconstruction Fiasco*. New York: Westview Press, 2005.

Pollack, Kenneth M. *Arabs at War: Military Effectiveness, 1948–1991*. Lincoln, NE: University of Nebraska Press, 2002.

Rabinovich, Itamar. *The War for Lebanon, 1970–1985*. Ithaca, NY: Cornell University Press, 1985.

Robertson, William G., ed. *Block by Block: The Challenges of Urban Operations*. Fort Leavenworth, KS: U.S. Army Command and General Staff College Press, 2003.

Robinett, Paul M. *Armor Command: The Personal Story of a Commander of the 13th Armored Regiment, of CCB, 1st Armored Division, and of the Armored School During World War II*. Washington, DC: McGregor & Werner, Inc., 1958.

Rogers, Hugh C.B. *Tanks in Battle*. London: Seeley Publishing, 1965.

Ross, G. MacLeod. *The Business of Tanks, 1933 to 1945*. Ilfracombe, England: Arthur H. Stockwell, 1976.

Schreier, Konrad F., Jr. *Standard Guide to US World War II Tanks & Artillery*. Iola, WI: Krause Publishing, 1994.

Simpkin, Richard. *Red Armour: An Examination of the Soviet Mobile Concept*. Washington, DC: Brassey's Defense Publisher, 1984.

________. *Tank Warfare: An Analysis of Soviet and NATO Tank Philosophy*. London: Brassey's Publishers, Ltd., 1979.

Smith, George W. *The Siege of Hue*. New York: Ballantine Books, 2000.

Spiller, Roger J., ed. *Combined Arms in Battle Since 1939*. Fort Leavenworth, KS: U.S. Army Command and General Staff College Press, 1992.

Starry, Donn A. *Armored Combat in Vietnam*. New York: Arno Press, 1980.

Stone, John. *The Tank Debate: Armour and the Anglo-American Military Tradition*. Amsterdam: Harwood Academic, 2000.

Sun Tzu. *Art of War*, trans. Samuel B. Griffith. New York: Oxford University Press, 1963.

Thompson, W. Scott and Donald D. Frizzell. *The Lessons of Vietnam*. New York: Crane, Russak & Company, 1977.

Tucker, Mike. *Among Warriors in Iraq: True Grit, Special Ops, and Raiding in Mosul and Fallujah*. Guilford, CN: The Lyon's Press, 2005.

Warr, Nicholas. *Phase Line Green: The Battle for Hue, 1968*. Annapolis, MD: Naval Institute Press, 1997.

Werstein, Irvin. *The Battle of Aachen.* New York: Thomas Y. Crowell Company, 1962.

West, Bing. *No True Glory: A Frontline Account of the Battle for Fallujah*. New York: Bantam Books, 2005.

Westmoreland, William C. *A Soldier Reports*. New York: Da Capo Press, 1976.

Whiting, Charles. *Bloody Aachen*. New York: PEI Books, Inc., 1976.

________. *Siegfried: The Nazis' Last Stand.* New York: Stein and Day Publishing, 1982.

________. *West Wall: The Battle for Hitler's Siegfried Line, September 1944–March 1945*. Conshohocken, PA: Combined Publishing, 2000.

Wright, Patrick. *Tank: The Progress of a Monstrous War Machine*. New York: Penguin Putnam, Inc., 2002.

Yazid Sayigh. *Arab Military Industry: Capability, Performance, and Impact*. London: Brassey's Defense Publishers, 1992.

Zaloga, Steven J. and James W. Loop. *Soviet Tanks and Combat Vehicles, 1946 to Present*. Dorset, England: Arms and Armour Press, 1987.

Zaloga, Steven J. *T-54, T-55, T-62*. New Territories, Hong Kong: Concord Publishing, 1992.

________. *T-64 and T-80*. New Territories, Hong Kong: Concord Publishing, 1992.

________. *The M2 Bradley Infantry Fighting Vehicle*. London, Osprey Publishing, 1986

Periodicals

Antal, John F. "Glimpse of Wars to Come: The Battle for Grozny." *Army* 49, no. 6 (June 1999): 28–34.

Betson, William R. "Tanks in Urban Combat." *Armor*, no. 4 (July–August 1992): 22–25.

Blank, Stephen. "Russia's Invasion of Chechnya: A Preliminary Assessment." Strategic Studies Institute, 1994.

Celestan, Major Gregory J. "Red Storm: The Russian Artillery in Chechnya," *Field Artillery* (January–February 1997): 42–45.

Chiarelli, Peter W. "Armor in Urban Terrain: the Critical Enabler." *Armor,* no. 2 (March–April 2005): 14–17.

Gabriel, Richard A. "Lessons of War: The IDF in Lebanon." *Military Review* (August 1984): 47–65.

Geibel, Adam. "Lessons of Urban Combat: Grozny 1994." *Infantry* 85, no. 6. (November–December 1995): 21–23.

Groves, Brigadier General John R. "Operations in Urban Environments." *Military Review* (July–August 1998): 7–12.

Milton, T.R. "Urban Operations: Future War." *Military Review* (February 1994): 37–46.
Mizrachi, Arie. "Israeli Artillery Tactics and Weapons: Lessons Learned in Combat." *Field Artillery*, no. 1 (February 1990): 7–10.
Peters, Ralph. "The Future of Armored Warfare." *Parameters* (Autumn 1997): 50–59.
Showalter, Dennis. "America's Armored Might." *World War II Magazine* 20, no. 1 (April 2005): 50–56.
Thomas, Timothy L. "Grozny 2000: Urban Combat Lessons Learned." *Military Review* (July–August 2000).
________. "The Battle of Grozny: Deadly Classroom for Urban Combat." *Parameters* (Summer 1999): 87–102.
Tiron, Roxana. "Heavy Armor Gains Clout in Urban Combat." *National Defense Magazine* (July 2004): 34–41.

Index